BOTANICAL RESEARCH AND PRACTICES

FLOWERING PLANTS

CLASSIFICATION, CHARACTERISTICS AND BREEDING

BOTANICAL RESEARCH AND PRACTICES

Additional books in this series can be found on Nova's website under the Series tab.

Additional E-books in this series can be found on Nova's website under the E-book tab.

BOTANICAL RESEARCH AND PRACTICES

FLOWERING PLANTS

CLASSIFICATION, CHARACTERISTICS AND BREEDING

JEREMY J. TELLSTONE
EDITOR

Nova Science Publishers, Inc.
New York

Copyright © 2011 by Nova Science Publishers, Inc.

All rights reserved. No part of this book may be reproduced, stored in a retrieval system or transmitted in any form or by any means: electronic, electrostatic, magnetic, tape, mechanical photocopying, recording or otherwise without the written permission of the Publisher.

For permission to use material from this book please contact us:
Telephone 631-231-7269; Fax 631-231-8175
Web Site: http://www.novapublishers.com

NOTICE TO THE READER

The Publisher has taken reasonable care in the preparation of this book, but makes no expressed or implied warranty of any kind and assumes no responsibility for any errors or omissions. No liability is assumed for incidental or consequential damages in connection with or arising out of information contained in this book. The Publisher shall not be liable for any special, consequential, or exemplary damages resulting, in whole or in part, from the readers' use of, or reliance upon, this material. Any parts of this book based on government reports are so indicated and copyright is claimed for those parts to the extent applicable to compilations of such works.

Independent verification should be sought for any data, advice or recommendations contained in this book. In addition, no responsibility is assumed by the publisher for any injury and/or damage to persons or property arising from any methods, products, instructions, ideas or otherwise contained in this publication.

This publication is designed to provide accurate and authoritative information with regard to the subject matter covered herein. It is sold with the clear understanding that the Publisher is not engaged in rendering legal or any other professional services. If legal or any other expert assistance is required, the services of a competent person should be sought. FROM A DECLARATION OF PARTICIPANTS JOINTLY ADOPTED BY A COMMITTEE OF THE AMERICAN BAR ASSOCIATION AND A COMMITTEE OF PUBLISHERS.

Additional color graphics may be available in the e-book version of this book.

Library of Congress Cataloging-in-Publication Data
Flowering plants : classification, characteristics and breeding / editor: Jeremy J. Tellstone.
    p. cm.
  Includes bibliographical references and index.
  ISBN 978-1-61324-653-5 (hardcover : alk. paper) 1. Angiosperms. 2. Angiosperms--Reproduction. 3. Angiosperms--Classification. I. Tellstone, Jeremy J.
  QK495.A1F575 2011
  582.13--dc22
                     2011015372

Published by Nova Science Publishers, Inc. † New York

CONTENTS

Preface		**vii**
Chapter 1	Response of Flower and Boll Development to Climatic Factors in Egyptian Cotton (*Gossypium barbadense*) *Zakaria M. Sawan*	**1**
Chapter 2	Risk Assessment of Inorganic and Organic Pollutants in Flowering Plants *Simona Dobrinas, Alina Daria Soceanu and Gabriela Stanciu*	**43**
Chapter 3	Seed Germination and the Secondary Metabolites, Plant Hormones *Mohammad Miransari*	**71**
Chapter 4	Rise of the Clones: Apomixis in Plant Breeding *Pablo Bolaños-Villegas, Saminathan Thangasamy and Guang-Yuh Jauh*	**97**
Chapter 5	Source/Sink Relations in Fruiting Cuttings of Grapevine (*Vitis Vinifera L.*) During the Inflorescence Development *Gaël Lebon, Florence Fontaine, Cédric Jacquard, Christophe Clément and Nathalie Vaillant-Gaveau*	**119**
Chapter 6	Improvement Strategies to Control Architecture and Flowering in Ornamental Plants Such as Azalea *M. Meijón, M. J. Cañal, R. Rodríguez, and I. Feito*	**137**

Chapter 7	The Evolution of Carnivory in Flowering Plants *Chris Thorogood*	**153**
Index		**179**

PREFACE

In this book, the authors gather and present topical research in the study of the classification, characteristics and breeding of flowering plants. Topics discussed include the response of flower and boll development to climatic factors in Egyptian cotton; risk assessment of inorganic and organic pollutants in flowering plants; seed germination and secondary metabolites and plant hormones; apomixis in plant breeding; source/sink relations in fruiting cuttings of grapevine during inflorescence development; controlling the architecture and flowering in ornamental azalea plants and the evolution of carnivory in flowering plants. (Imprint: Nova)

Chapter 1 - Fruiting of cotton plant is determined and influenced by cultivars, climatic conditions, management practices and pests. An understanding of the flowering and boll retention patterns of cotton cultivars can contribute to more efficient and economical crop management. The objective of this investigation was to study the effect of various climatic factors on flower and boll production, and also, the nature of its effects prevailing prior and subsequent to either flowering or boll setting on flower and boll production and retention in Egyptian cotton. This could be used in formulating advanced prediction of the effect of certain climatic conditions on the production of Egyptian cotton. Also, the study focused on four equal periods during the development of flower and bolls stage to study the response of these characters to climatic factors during these periods and to determine the most representative period corresponding to the overall crop pattern. Further, the study predicting effects of climatic factors during convenient intervals (in days) on cotton flower and boll production compared with daily observations to find the optimum interval. Evaporation, sunshine duration, humidity, surface soil temperature at 1800 h, and maximum air temperature, are the important climatic factors that significantly affects flower and boll production. Evaporation; minimum humidity and sunshine duration were the most effective climatic factors during preceding and succeeding periods on boll production and retention. There was a negative correlation between flower and boll production and either evaporation or sunshine duration, while that correlation with minimum humidity was positive. The fourth quarter period of the production stage was the most appropriate and usable time to collect data for determining efficient prediction equations for cotton production. Evaporation, humidity and temperature were the principle climatic factors that governed cotton flower and boll production during the fourth quarter. The five day interval was found to be adequately and sensibly related to yield parameters than other intervals and was closest to the daily observations. Evaporation was found to be the most important climatic variable affecting flower and boll production, followed by humidity. An accurate weather forecast 5-7 days in advance would provide an opportunity to avoid adverse effects of climatic factors on cotton production through utilizing proper cultural practices which would limit and control their negative effects.

Chapter 2 - The first step in the risk assessment process is to identify potential health effects that may occur from different types of pollutants exposure. Humans are exposed to pollutants by different routes of exposure such as inhalation, ingestion and dermal contact. Heavy metals are very harmful because of their non-biodegradable nature, long biological half-lives and their potential to be accumulated in different body parts. Organochlorine pesticides (OCPs) were used for the first time in Romania in 1948. Since 1988 these kinds of products were banned or restricted in Romania and in the present only chlorinated insecticides on the base of lindane are used for seeds treatment in Romania, but this substance is not included in the Stockholm Convention list on Persistent Organic Pollutants. Human exposure to polycyclic aromatic hydrocarbons (PAHs) can occur through different environmental pathways, including internal absorption through food and water consumption. Human exposure to PAHs is 88-98% connected with food (in 5% with food of plant origin). In this study levels of heavy metals (Pb, Cd, Cu, Zn, Mn, Fe), OCPs (lindane, p,p'- DDT, p,p'- DDE, p,p'- DDD, HCB, aldrin, dieldrin, endrin and hepthaclor) and PAHs (Np, Acy, Ace, F, Ph, An, Fl, Py, B[a]An, Chry, B[k]Fl, B[a]Py, B[ghi]P, dB[a,h]An, I[1,2,3-cd]Py) in flowering plants from Solanaceae and Rosaceae families were investigated. Studied plants at various growing stages were: tomato and bell pepper plants at 5, 15 and 35 cm (samples were taken from roots, stems, leaves and fruits) and peach, nectarine, apricot, cherry, sour cherry, apple and quince trees (green, almost ripe and ripe fruits) from Romania's urban and rural areas. For metals determination was used a AA6200 Schimadzu FAAS. Analysis of PAHs was carried with a HP 5890/5972 GC-MS system and analysis of OCPs was carried with a HP 5890 gas chromatograph equipped with an electron capture detector. Repeatability of methods, expressed as the relative standard deviation, was lower than 7.5% while recoveries were in the range of 96–99%. The estimated provisional tolerable weekly intake (PTWI) of all studied metals was calculated. All pesticides and PAHs concentrations of analyzed flowering plants were compared with values imposed by European Communities regulations. According to ANOVA test, statistically significant differences were found between samples from urban and rural areas, respectively, among samples within each botanical origin. Data analysis of health risk estimates indicated that analyzed persistent organic pollutants do not pose a direct hazard to human health.

Chapter 3 - Seed germination is interesting and complicated. It is controlled by different mechanisms and is necessary for the growth and development of embryo, which eventually produces a new plant. Under unfavorable conditions seeds become dormant to maintain their germination ability. However, when the conditions are favorable seeds can germinate. There are different parameters controlling seed germination and dormancy, among which plant hormones are the most important ones. Plant hormones are produced by both plants and soil bacteria and control different processes related to seed growth and development. In this review article some of the most recent findings regarding seed germination and dormancy as well as the significance of plant hormones including abscisic acid, ethylene, gibberellins, auxin, cytokinins and brassinosteroids with reference to proteomic and molecular biology studies on such phenomena are discussed. In the conclusion some idea for the future research is expressed.

Chapter 4 - In flowering plants, the transfer of traits from one generation to another is accomplished by fertilization of the female gametophyte with sperm cells delivered by the pollen tube and subsequent reassortment of traits (alleles) in the developing progeny. In nature, DNA recombination and segregation of traits in the progeny prevent the accumulation of deleterious genes and loss of fitness; however, for breeding purposes, fixing superior trait combinations (genotypes) and circumventing sexual reproduction is advantageous. The formation of sexual seed without fertilization of the egg is called apomixis and is considered the "holy grail" of plant breeding because top-performing varieties can be reproduced indefinitely without changes in the genotypes themselves or their expression patterns. Apomixis is a dominant trait and consists of several processes working in tandem; separately, each process is detrimental for plant reproduction, but as a single unit, they allow development of embryos and endosperm from unfertilized eggs. The 3 processes of apomixis include 1) apomeiosis, or cell division without DNA recombination in pollen and eggs; 2) parthenogenesis, autonomous development of eggs into fully formed embryos; and 3) stable development of the endosperm, the part of the seed needed for the embryo to grow. Unfortunately, full expression and transmission of apomixis is affected by DNA recombination; therefore, introgression of the trait from wild relatives into commercial varieties is extremely difficult. In this review, we report on genes that have been identified to regulate each step of apomixis and discuss strategies to allow full transmission of the trait with tools from molecular biology.

Chapter 5 - In crops, flower and fruit abscission induce significant yield loss. Dependant on the sink strength of each organ, carbohydrate supply to reproductive structures is a putative physiological cause of this phenomenon. In the present work, the effect of defoliation, as an agent of source/sink modifications, was studied in grapevine (*Vitis vinifera* L.) using fruiting cuttings. In control cuttings only four leaves were allowed to develop. In order to modify artificially source/sink interactions, another set of cuttings was systematically defoliated (0L) whereas the last set was characterized by the free development of all the leaves (AL). The results show that developing leaves have a growth-depressing effect on inflorescences and roots, showing their stronger sink strength. Besides, 0L and AL treatments modify carbohydrate concentrations in all organs of both cvs., especially in inflorescences at crucial steps of male and female organ formation (meiosis). However the variations did not follow the same pathway in each cv., likely due to different carbohydrate metabolism. Whatever, carbohydrate levels in cuttings inflorescences were different than those of vineyard inflorescences.

Chapter 6 - The evergreen azalea (*Rhododendron* L. sp) is a woody shrub widely used in gardens and in fact the genus *Rhododendron* is among the most popular landscape plants in Europe and North America. It is also sold as a greenhouse-grown potted plant, marketed in flower, for decorative indoor use. The genus contains approximately 1000 described species and thousands of commercial hybrids with new cultivars entering the market every year. Plants are grown as densely branched shrubs in containers of various sizes. However, many cultivars show a very strong growth tendency and an irregular shape.

Chapter 7 - Carnivorous plants are extremely derived and many species have evolved distinct modifications in gross vegetative structures to complement their derived mode of nutrition. This divergence in morphological features has deprived taxonomic scientists of traits with which to delineate the evolutionary relationships of carnivorous plant families with their non-carnivorous ancestors. Carnivorous plants have long aroused the interest of botanists, and were used as model systems for physiological studies of movement and secretion in plants, pioneered by Darwin (1875). Studies on the evolution of carnivorous plants were depauperate in the literature for most of the 20th Century, and their origins remained largely speculative and unsubstantiated. However, advances in molecular technology and the advent of gene sequencing and molecular phylogenetics towards the end of the 20th Century have vastly improved our understanding of the evolution of flowering plants, including the origins of carnivory.

In: Flowering Plants
Editor: Jeremy J. Tellstone

ISBN: 978-1-61324-653-5
© 2011 Nova Science Publishers, Inc.

Chapter 1

RESPONSE OF FLOWER AND BOLL DEVELOPMENT TO CLIMATIC FACTORS IN EGYPTIAN COTTON (*GOSSYPIUM BARBADENSE*)

Zakaria M. Sawan[*]
Cotton Research Institute, Agricultural Research Center,
Ministry of Agriculture & Land Reclamation,
9 Gamaa Street, 12619, Giza, Egypt

ABSTRACT

Fruiting of cotton plant is determined and influenced by cultivars, climatic conditions, management practices and pests. An understanding of the flowering and boll retention patterns of cotton cultivars can contribute to more efficient and economical crop management. The objective of this investigation was to study the effect of various climatic factors on flower and boll production, and also, the nature of its effects prevailing prior and subsequent to either flowering or boll setting on flower and boll production and retention in Egyptian cotton. This could be used in formulating advanced prediction of the effect of certain climatic conditions on the production of Egyptian cotton. Also, the study focused on four equal periods during the development of flower and bolls stage to study the response of these characters to climatic factors during

---

[*] Correspondance Address: zmsawan@hotmail.com

these periods and to determine the most representative period corresponding to the overall crop pattern. Further, the study predicting effects of climatic factors during convenient intervals (in days) on cotton flower and boll production compared with daily observations to find the optimum interval. Evaporation, sunshine duration, humidity, surface soil temperature at 1800 h, and maximum air temperature, are the important climatic factors that significantly affects flower and boll production. Evaporation; minimum humidity and sunshine duration were the most effective climatic factors during preceding and succeeding periods on boll production and retention. There was a negative correlation between flower and boll production and either evaporation or sunshine duration, while that correlation with minimum humidity was positive. The fourth quarter period of the production stage was the most appropriate and usable time to collect data for determining efficient prediction equations for cotton production. Evaporation, humidity and temperature were the principle climatic factors that governed cotton flower and boll production during the fourth quarter. The five day interval was found to be adequately and sensibly related to yield parameters than other intervals and was closest to the daily observations. Evaporation was found to be the most important climatic variable affecting flower and boll production, followed by humidity. An accurate weather forecast 5-7 days in advance would provide an opportunity to avoid adverse effects of climatic factors on cotton production through utilizing proper cultural practices which would limit and control their negative effects.

ABBREVIATIONS

ET   Evapotranspiration
ET_{max} Maximum Evapotranspiration
PGR   Plant growth regulators

1. INTRODUCTION

Agronomists and crop production specialists are often unable to determine the effect of various environmental factors on crop growth, development and yield. This results from difficulty in measuring plant response to different environmental conditions and from the covariance and interactions of environmental factors in the field. A method is needed to provide quantitative information regarding plant response to weather, soil and management conditions. Modeling and simulation of plant response to weather and soil conditions could possibly be a useful tool for identifying factors limiting plant growth in complex environments. Because the climatic factors are largely uncontrolled, thus variables that influence flower and boll production must be quantitatively evaluated if we want to explain adequately the effects of any climatic variable on cotton flower and boll production. So it is important to determine the most important climatic factors on increasing or decreasing cotton production and its magnitude. Understanding of the impact may help the physiologist to determine possible control of flowering mechanism in cotton plant.

General circulation models (GCMs) project increases of the earth's surface air temperatures and other climate changes in the middle or latter part of the 21st century, and therefore crops such as cotton (*Gossypium hirsutum* L.) will be grown in a much different environment than today (Reddy et al. 2002). Therefore, if global warming occurs as projected, fiber production in the future environment will be reduced and breeding extreme temperature tolerant cultivars will be necessary to sustain cotton production in the US. Cultural practices such as earlier planting may be used to avoid the flowering of cotton in the high temperatures that occur during mid to late summer (Reddy et al. 2002). Schrader et al. (2004) stated that high temperatures that plants are likely to experience inhibit photosynthesis.

El-Zik (1980) stated that many factors, such as length of the growing season, climate (including solar radiation, temperature, light, wind, rainfall, and dew), variety, availability of nutrients and soil moisture, pests and cultural practices affect cotton growth. The balance between vegetative and reproductive development can be influenced by soil fertility, soil moisture, cloudy weather, spacing and perhaps other factors such as temperature and humidity (Guinn, 1982).

In Texas, Guo et al. (1994) found that plant growth and yield of the cotton cv. DPL-50 (Upland cotton) were less in a humid area than in an arid area with low humidity. Under arid conditions, high vapor pressure deficit resulted in a high transpiration rates, low leaf water potential and lower leaf temperatures. Reddy et al. (1995), in growth chamber experiments found that Pima cotton cv. S-6 produced less total biomass at 35.5°C than at 26.9°C and no bolls were produced at 40°C. Miller et al. (1996) ran multiple regressions between rainfall, temperature and yield data gathered from 1968-92 and found that in most cases less than 50% of the yield variation for non-irrigated cotton can be explained by a combination of weather factors. The other 50% of yield variation is subject to management.

Gipson and Joham (1968) reported that cool temperatures (< 20°C) at night hinder boll development. Fisher (1975) found that high temperatures can cause male sterility in cotton flowers, and could have caused increased boll shedding in the late fruiting season. Zhao (1981) indicated that temperature was the main climatic factor affecting cotton production and 20-30°C was the optimum temperature for cotton growth. McKinion et al. (1991) found that Pima cotton (*G. barbadense* L.) plants did not produce fruiting branches or reproductive organs above 40/32°C (day/night) conditions. Hodges et al. (1993) found that the optimum temperature for cotton stem and leaf growth, seedling and fruiting was almost 30°C. Fruit retention decreased rapidly as the time of exposure to 40°C increased.

Seedling establishment and development of a canopy structure capable of interception light and supporting fruit or grain production is fundamental to crop production and is highly temperature dependent. As more is learned about crop production, more specific information is needed. We would like to stimulate growth and developmental processes to accurately predict plant responses to weather, soil, and management-influenced conditions. We need specific rates of development and duration of leaf growth to do that because leaf area is fundamental to light interception (Reddy et al. 1993). There appears to be a difference of opinions regarding the effects of temperature on boll setting, perhaps because cultivars differ and the way the experiments were conduced could be also different. Burke et al. (1988) has defined the optimum temperature range for biochemical and metabolic activities of plants (a temperature range that permits normal enzyme functions in plants) as the thermal kinetic window (TKW). Temperature above or below the TKW results in stress that limits growth and yield. The TKW for cotton growth is 23.5 to 32°C, with an optimum temperature of 28°C. Biomass production is directly related to the amount of time that foliage temperature is within the TKW.

Reddy et al. (1999) indicated that, an experiment was conducted in naturally lit plant growth chambers to determine the influence of temperature and atmospheric [CO_2] on cotton (*Gossypium hirsutum* cv. DPL-51) boll and fiber growth parameters. Boll size and maturation periods decreased as temperature increased. Boll growth increased with temperature to 25°C and then declined at the highest temperature.

Zhou et al. (2000) indicated that sunshine hours is the key meteorological factor influencing the wheat-cotton cropping pattern and position of the bolls, while temperature had an important function on upper (node 7 to 9) and top (node 10) bolls, especially for double cropping patterns with early maturing varieties.

Barbour and Farquhar (2000), in greenhouse pot trials, cotton cv. CS50 plants were grown at 43 or 76% RH and sprayed daily with abscisic acid (ABA) at 1 x 10^{-5}, 1 x 10^{-4} or 1 x 10^{-3} or distilled water. Plants grown at the lower humidity had higher transpiration rates, lower leaf temperatures and lower stomatal conductance. Plant biomass was also reduced at the lower humidity. Within each humidity environment, increasing ABA concentration generally reduced stomatal conductance, evaporation rates, superficial leaf density and plant biomass, and increased leaf temperature and specific leaf area. Oosterhuis (1999) stated that, high day temperatures and water stress result in low boll weights and reduced cotton yields. Yuan et al. (2002) pointed out that, correlation analysis of meteorological data (1987-95) showed that under the semi-arid ecological conditions on the Loess Plateau, dryness is the main factor limiting the development of cotton production. Evaporation is the crux that influences the cotton yield. It is suggested that a comprehensive management system should center on the reduction of evaporation, and promoting of water utilization efficiency.

The objectives of this investigation were to study: 1- The effect of various climatic factors on the overall flower and boll production in Egyptian cotton. This could pave the way for formulating advanced predictions as to the effect of certain climatic conditions on cotton production of Egyptian cotton. It would be useful to minimize the deleterious effects of the factors through utilizing proper cultural practices which would limit and control their negative effects, and this will lead to an increase in cotton yield. 2- The work reported here raises a question regarding the nature of the effects of specific climatic factors during both pre- and post-anthesis periods on boll production and retention. Hence, the objective of this investigation was to study and collect information about the nature of the relationship between various climatic factors and cotton boll development and the 15-day period both prior to and after initiation of individual bolls of field-grown cotton plants in Egypt. As a matter of fact an understanding of the relationships between climatic factors, flowering and boll-retention patterns of the cotton plants during these periods might allow a direct external intervention that can help cotton growth and production. 3- Also, this study investigated the relationship between climatic factors and production of flowers and bolls obtained during the development periods of the flowering and boll stage, and to determine the most representative period corresponding to the overall crop pattern. 4- Further, provide information on cotton responses to the climatic factors during a long period (about two months) of flower and boll production. Correlation of condensed data, i.e., optimum number of collecting days of sound climatic elements mostly related to cotton production to give realistic or near realistic results.

2. DATA AND METHODS

Two uniform field trials were conducted at the experimental farm of the Agricultural Research Center, Ministry of Agriculture, Giza, Egypt (30° N, 31°: 28'E), using the cotton cultivar Giza 75 (*Gossypium barbadense* L.) in 2 successive seasons (I and II). The soil texture was a clay loam, with an alluvial substratum, (pH = 8.07, 42.13% clay, 27.35% silt, 22.54% fine sand, 3.22% coarse sand, 2.94% calcium carbonate and 1.70% organic matter).

Total water consumed during each of the 2 growing seasons supplied by surface irrigation was about 6000-m^3 h^{-1}. The criteria used for watering the crop depended on soil water status, where irrigation was applied when soil water content reached about 35% of field capacity. In Season I, the field was irrigated on 15 March (at planting), 8 April (first irrigation), 29 April, 17 May, 31 May, 14 June, 1 July, 16 July and 12 August. In Season II, the field was irrigated on 23 March (planting date), 20 April (first irrigation), 8 May, 22 May, 1 June, 18 June, 3 July, 20 July, 7 August and 28 August. Techniques normally used for growing cotton in Egypt were followed. Each experimental plot contained 13 to15 ridges to facilitate proper surface irrigation. Ridge width was 60 cm and its length was 4 m. Seeds were sown on 15 and 23 March in Seasons I and II, respectively, in hills 20 cm apart on one side of the ridge. Seedlings were thinned to 2 plants per hill 6 wk after planting, resulting in a plant density of about 166 000 plants ha^{-1}. Phosphorus fertilizer was applied at a rate of 54 kg P_2O_5 ha^{-1} as calcium superphosphate during land preparation. Potassium fertilizer was applied at a rate of 57 kg K_2O ha^{-1} as potassium sulphate before the first irrigation. Nitrogen fertilizer was applied at a rate of 144 kg N ha^{-1} as ammonium nitrate with lime 2 equal doses: the first applied after thinning just before the second irrigation and the other applied before the third irrigation. Rates of phosphorus, potassium, and nitrogen fertilizer were the same in both years.

After thinning, 261 and 358 plants were randomly selected (precaution of border effect was taken into consideration by discarding the cotton plants in the first and last 2 hills each ridge) from 9 and 11 inner ridges of the plot in Seasons I, and II respectively. Flowers on all selected plants were tagged in order to count and record the number of open flowers, and set bolls on a daily basis. The flowering season commenced on the date of the first flower appearance and continued until the end of flowering season (31 August), which would give sound bolls at the end of the handpicking season (20 October). Each flower was tagged according to date of appearance on the selected plants. In Season I, the flowering period extended from 17 June to 31 August, whereas in Season II, the flowering period was from 21 June to 31 August. Flowers produced after 31 August, are not expected to form sound harvestable bolls, and therefore were not taken into account. As a rule, observations were recorded when the number of flowers on a given day was at least 5 flowers found for a population of 100 plants and this continued for at least 5 consecutive days. This rule omitted 8 observations in the first season and 10 observations in the second season. So the number of observations (n) was 68 (23 June-29 August) and 62 (29 June-29 August) for the two seasons, respectively.

For statistical analysis, the following data of the dependent variables were determined: (1) daily number of tagged flowers separately counted each day on all selected uniform plants, (2) number of retained bolls obtained from the total daily tagged flower on all selected plants at harvest; and (3) percentage of boll retention ([number of retained bolls obtained from the total number of daily tagged flowers in all selected plants at harvest]/[daily number of tagged flowers on each day in all selected plants] x 100).

The climatic factors (independent variables) considered were daily data of: maximum air temperature (°C, X_1); minimum air temperature (°C, X_2); maximum-minimum air temperature (diurnal temperature range) (°C, X_3); evaporation (expressed as Piche evaporation) (mm day^{-1}, X_4); surface soil temperature, grass temperature or green cover temperature at 0600 h (°C, X_5) and 1800 h (°C, X_6); sunshine duration (h day^{-1}, X_7); maximum humidity (%, X_8), minimum humidity (%, X_9) and wind speed (m s^{-1}, X_{10}) (in season II only). The source of the climatic data was the Agricultural Meteorological Station of the Agricultural Research Station, Agricultural Research Center, Giza, Egypt. No rainfall occurred during the 2 growing seasons.

2.1. Statistical Analysis

Statistical analysis was conducted using the procedures outlined in the general linear model (GLM, SAS Institute, Inc. 1985).

3. RESULTS AND DISCUSSION

3.1. Response of Flower and Boll Development to Climate Factors on the Anthesis Day

Daily number of flowers and number of bolls per plant which survived to maturity (dependent variables) during the production stage of the two seasons (68 days and 62 days in the first and the second seasons, respectively) are graphically illustrated in Figures 1 and 2 (Sawan et al. 2002a). The flower- and boll-curves reached their peaks during the middle two weeks of August, and then descended steadily till the end of the season. Specific differences in the shape of these curves in the two seasons may be due to the growth-reactions of environment, where climatic factors (Table 1) (Sawan et al. 2005) represent an important part of the environment effects (Miller et al. 1996).

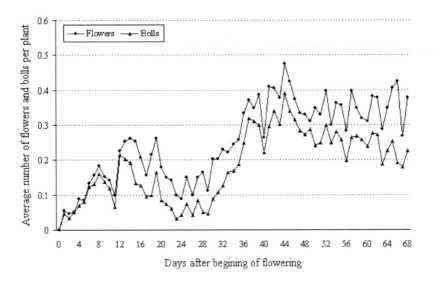

Figure 1. Daily number of flowers and bolls during the production stage (68 days) in the first season (I) for the Egyptian cotton cultivar Giza 75 (*Gossypium barbadense* L.) grown in uniform field trial at the experimental farm of the Agricultural Research Centre, Giza (30°N, 31°:28'E), Egypt. The soil texture was a clay loam, with an alluvial substratum, (pH = 8.07). Total water consumptive use during the growing season supplied by surface irrigation was about 6000 m^{3}ha^{-1}. No rainfall occurred during the growing season. The sampling size was 261 plants (Sawan et al., 2005).

3.1.1. Correlation Estimates

Results of correlation coefficients [correlation and regression analyses were computed, according to Draper and Smith (1966) by means of the computer program SAS package (1985).] between the initial group of independent variables and each of flower and boll production in the first and second seasons and the combined data of the two seasons are shown in Table 2 (Sawan et al. 2002a).

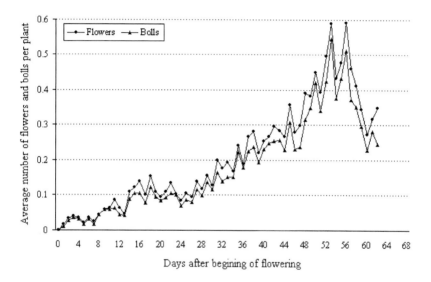

Figure 2. Daily number of flowers and bolls during the production stage (62 days) in the second season (II) for the Egyptian cotton cultivar Giza 75 (*Gossypium barbadense* L.) grown in uniform field trial at the experimental farm of the Agricultural Research Centre, Giza (30°N, 31°:28'E), Egypt. The soil texture was a clay loam, with an alluvial substratum, (pH = 8.07). Total water consumptive use during the growing season supplied by surface irrigation was about 6000 m^{3}ha^{-1}. No rainfall occurred during the growing season. The sampling size was 358 plants (Sawan et al., 2005).

The correlation values indicate clearly that evaporation is the most important climatic factor affecting flower and boll production as it showed the highest correlation value. Also, this factor had a significant negative relationship with flower and boll production. Sunshine duration showed a significant negative relation with fruit production except for boll production in the first season, which was not significant. Maximum air temperature, temperature magnitude, and surface soil temperature at 1800 h, were also negatively correlated with flower and boll production in the second season and the combined data of the two seasons. Minimum humidity in the second season, the combined data of the two seasons, and maximum humidity in the first season were positively and highly correlated with flower and boll production. Minimum air temperature and surface soil temperature at 0600 h showed low and insignificant correlation to flower and boll production (Sawan et al. 2002a).

Table 1. Mean, standard deviation, maximum and minimum values of the climatic factors during the flower and boll stage (initial time) and the 15 days prior to flowering or subsequent to boll setting for I and II season at Giza, Egypt

Climatic factors	First season*				Second season**			
	Mean	S.D.	Max.	Min.	Mean	S.D.	Max.	Min.
Max temp [°C] (X_1)	34.1	1.2	44.0	31.0	33.8	1.2	38.8	30.6
Min temp [°C] (X_2)	21.5	1.0	24.5	18.6	21.4	0.9	24.3	18.4
Max-Min temp [°C] (X_3)$^\bullet$	12.6	1.1	20.9	9.4	12.4	1.3	17.6	8.5
Evapor [mm d^{-1}](X_4)	10.6	1.6	16.4	7.6	6.0	0.7	9.8	4.1
0600 h temp [°C] (X_5)	17.5	1.1	21.5	13.9	17.6	1.2	22.4	13.3
1800 h temp [°C] (X_6)	24.2	1.9	32.3	19.6	23.7	1.1	27.4	20.6
Sunshine [h d^{-1}] (X_7)	11.7	0.8	12.9	9.9	11.7	0.4	13.0	10.3
Max hum [%] (X_8)	85.6	3.3	96.0	62.0	72.9	3.8	84.0	51.0
Min hum [%] (X_9)	30.2	5.2	45.0	11.0	39.1	5.0	52.0	23.0
Wind speed [m s^{-1}] (X_{10})	ND	ND	ND	ND	4.6	0.9	7.8	2.2

*Flower and boll stage (68 days, from 23 June through 29 August).
**Flower and boll stage (62 days, from 29 June through 29 August).
$^\bullet$ diurnal temperature range.
ND not determined
(Sawan et al., 2005)

The negative relationship between evaporation with flower and boll production, means that high evaporation rate significantly reduces cotton flower and boll production. This may be due to greater plant water deficits when evaporation increases. Also, the negative relation between each of maximum temperature, temperature magnitude, soil surface temperature at 1800 h, or sunshine duration, with flower and boll production revealed that the increase in the values of these factors had a detrimental effect upon fruit production in Egyptian cotton. On the other hand, there was a positive correlation between each of maximum or minimum humidity with flower and boll production (Sawan et al. 2002a).

Table 2. Simple correlation values for the relationships between the independent variables and the studied dependent variable

Independent variables		Dependent variable					
		First season		Second season		Combined data	
(Climatic factors)		Flower	Boll	Flower	Boll	Flower	Boll
Max Temp [°C]	(X_1)	-0.07	-0.03	-0.42**	-0.42**	-0.27**	-0.26**
Min Temp [°C]	(X_2)	-0.06	-0.07	0.00	0.02	-0.03	-0.02
Max-Min Temp [°C]	(X_3)	-0.03	-0.01	-0.36**	-0.37**	-0.25**	-0.24**
Evapor [mm/d]	(X_4)	-0.56**	-0.53**	-0.61**	-0.59**	-0.40**	-0.48**
0600 h Temp [°C]	(X_5)	-0.01	-0.06	-0.14	-0.13	-0.09	-0.09
1800 h Temp [°C]	(X_6)	-0.20	-0.16	-0.37**	-0.36**	-0.27**	-0.25**
Sunshine [h/d]	(X_7)	-0.25*	-0.14	-0.37**	-0.36**	-0.31**	-0.25**
Max Hum [%]	(X_8)	0.40**	0.37**	0.01	0.01	0.04	-0.06
Min Hum [%]	(X_9)	0.14	0.10	0.45**	0.46**	0.33**	0.39**
Wind speed [m/s]	(X_{10})	ND	ND	-0.06	-0.04	ND	ND

ND not determined

* $P < 0.05$; ** $P < 0.01$.

(Sawan et al., 2002a)

Results obtained from the production stage of each season individually, and the combined data of the two seasons, indicate that relationships of some climatic variables with the dependent variables varied markedly from one season to another. This may be due to the differences between climatic factors in the two seasons as illustrated by the ranges and means shown in Table 1. For example, maximum temperature, minimum humidity and surface soil temperature at 1800 h did not show significant relations in the first season, while that trend differed in the second season. The effect of maximum humidity varied markedly from the first season to the second. While it was significantly correlated with the dependent variables in the first season, the inverse pattern was true in the second season. This diverse effect may be due to the differences in the mean values of this factor in the two seasons; where it was, on average, about 86% in the first season, and on average about 72% in the second season, as shown in Table 1 (Sawan et al. 2005).

These results indicate that evaporation is the most effective and consistent climatic factor affecting boll production. As the sign of the relationship was negative, this means that an increase in evaporation would cause a significant reduction in boll number. Thus, applying specific treatments such as an additional irrigation, and use of plant growth regulators, which would decrease the deleterious effect of evaporation after boll formation and hence contribute to an increase in cotton boll production and retention, and the consequence is an increase in cotton yield. In this connection, Moseley et al. (1994) stated that methanol has been reported to increase water use efficiency, growth and development of C_3 plants in arid conditions, under intense sunlight. In field trials cotton cv. DPL-50 (*Gossypium hirsutum*), was sprayed with a nutrient solution (1.33 lb N + 0.27 lb Fe + 0.27 lb Zn acre^{-1}) or 30% methanol solution at a rate of 20 gallons acre^{-1}, or sprayed with both the nutrient solution and methanol under two soil moisture regimes (irrigated and dry land). The foliar spray treatments were applied 6 times during the growing season beginning at first bloom. They found that irrigation (a total of 4.5 inches applied in July) increased lint yield across foliar spray treatments by 18%. Zhao and Oosterhuis (1997) found that in a growth chamber when cotton (*Gossypium hirsutum* cv. Stoneville 506) plants were treated with the plant growth regulator PGR-IV (gibberellic acid, IBA and a proprietary fermentation broth) under water deficit stress had significantly higher dry weights of roots and floral buds than the untreated water-stressed plants. They concluded that PGR-IV can partially alleviate the detrimental effects of water stress on photosynthesis and dry matter accumulation and improves the growth and nutrient absorption of growth chamber-grown cotton plants. Meek et al. (1999) in a field experiment in Arkansas found that application of 3 or 6 kg glycine betaine (PGR) ha^{-1}, to cotton plants has the potential to increase yield in cotton exposed to mild water stress.

3.1.2. Multiple Linear Regression Equation

By means of the multiple linear regression analyses, fitting predictive equations (having good fit) were computed for flower and boll production per plant using selected significant factors from the nine climatic variables studied in this investigation. Wind speed evaluated during the second season had no influence on the dependent variables. The equations obtained for each of the two dependent variables, i.e. number of flowers (Y_1) and bolls per plant (Y_2)

Response of Flower and Boll Development to Climatic Factors ...   13

in each season and for combined data from the two seasons(Sawan et al., 2002a) are as follows:

First Season: (n = 68)
Y_1 = 21.691 - 1.968 X_4 - 0.241 X_7 + 0.216 X_8, R = 0.608** and R^2 = 0.3697, while R^2 for all studied variables was 0.4022 (Table 2)
Y_2 = 15.434 - 1.633 X_4 + 0.159 X_8, R = 0.589** and R^2 = 0.3469 and R^2 for all studied variables was 0.3843 (Table 2).

Second Season: (n = 62)
Y_1 = 77.436 - 0.163 X_1 - 2.861 X_4 - 1.178 X_7 + 0.269 X_9, R = 0.644**, R^2 = 0.4147
Y_2 = 66.281 - 0.227X_1 - 3.315X_4 - 2.897X_7 + 0.196X_9, R = 0.629**, R^2 = 0.3956.
In addition, R^2 for all studied variables was 0.4503 and 0.4287 for Y_1 and Y_2 equations respectively (Table 2).
Combined data for the two seasons: (n = 130)

Y_1 = 68.143 - 0.827 X_4 - 1.190 X_6 - 2.718 X_7 + 0.512 X_9, R = 0.613**, R^2 = 0.3758
Y_2 = 52.785 - 0.997 X_4 - 0.836 X_6 - 1.675 X_7 + 0.426 X_9, R = 0.569**, R^2 = 0.3552
While R^2 for all studied variables was 0.4073 for Y_1 and 0.3790 for Y_2.

Three climatic factors, i.e. minimum air temperature, surface soil temperature at 0600 h, and wind speed were not included in the equations since they had very little effect on production of cotton flowers and bolls (Sawan et al. 2002a). The sign of the partial regression coefficient for an independent variable (climatic factor) indicates its effect on the production value of the dependent variable (flowers or bolls). This means that high rates of humidity and/or low values of evaporation will increase fruit production (Sawan et al. 2002a).

3.1.3. *Contribution of Selected Climatic Factors to Variations in the Dependent Variable*
Relative contributions (RC %) for each of the selected climatic factors to variation in flower and boll production is summarized in Table 3 (Sawan et al. 2002a). Results in this table indicate that evaporation was the most important climatic factor affecting flower and boll production in Egyptian cotton.

Sunshine duration is the second climatic factor of importance affecting production of flowers and bolls. Humidity and temperature at 1800 h were contributing factors but less than evaporation and sunshine duration day^{-1}. Maximum temperature made a contribution but less than the other affecting factors (Sawan et al. 2002a).

Table 3. Selected factors and their relative contribution to variations of flower and boll production

Selected factors		Flower production			Boll production		
		* R.C. (%)			R.C. (%)		
		First season	Second season	Combined data	First season	Second season	Combined data
Max Temp [°C]	(X$_1$)	-	5.92	-	-	5.03	-
Evapor [mm/d]	(X$_4$)	19.08	23.45	16.06	23.04	22.39	22.89
1800 h Temp [°C]	(X$_6$)	-	-	5.83	-	-	2.52
Sunshine [h/d]	(X$_7$)	9.43	7.77	8.31	11.65	7.88	5.47
Max Hum [%]	(X$_8$)	8.46	-	-	-	-	-
Min Hum [%]	(X$_9$)	-	4.37	7.38	-	4.26	4.64
** R^2% for selected factors		36.97	41.47	37.58	34.69	39.56	35.52
R^2% for factors studied		40.22	45.03	40.73	38.43	42.87	37.90
R^2% for factors deleted		3.25	3.56	3.15	3.74	3.31	2.38

* R.C.% = Relative contribution of each of the selected independent variables to variations of the dependent variable.
** R^2 % = Coefficient of determination in percentage form.
(Sawan et al., 2002a)

Evaporation showed the highest contribution to the variation in both flower and boll production (Sawan et al. 2002a). This finding can, however, be explained in the light of results found by Ward and Bunce (1986) in sunflower (*Helianthus annuus*). They stated that decreases of humidity at both leaf surfaces reduced photosynthetic rate of the whole leaf for plants grown under a moderate temperature and medium light level. Kaur and Singh (1992) found in cotton that flower number was decreased by water stress, particularly when applied at flowering. Seed cotton yield was about halved by water stress applied at flowering, slightly decreased by stress at boll formation, and not significantly affected by stress in the vegetative stage (6-7 weeks after sowing). Orgaz et al. (1992) in field experiments at Cordoba, SW Spain, grew cotton cultivars Acala SJ-C1, GC-510, Coker-310 and Jean at evapotranspiration (ET) levels ranging from 40 to 100% of maximum ET (ET_{max}) which were generated with sprinkler line irrigation. The water production function of Jean cultivar was linear; seed yield was 5.30 t ha^{-1} at ET_{max} (820 mm). In contrast, the production function of the three other cultivars was linear up to 85% of ET_{max}, but leveled off as ET approached ET_{max} (830 mm) because a fraction of the set bolls did not open by harvest at high ET levels. These authors concluded that it is possible to define an optimum ET deficit for cotton based on cultivar earliness, growing-season length, and availability of irrigation water.

The negative relationship between sunshine duration and cotton production (Sawan et al. 2002a) may be due to the fact that the species of *Gossypium* spp. used is known to be a short day plant (Hearn and Constable 1984). So, an increase of sunshine duration above that needed for cotton plant growth will decrease flower and boll production. Bhatt (1977) found that, exposure to day light over 14 hours and high day temperature, individually or in combination, delayed flowering of the Upland cotton J 34.

Reddy et al. (1996) observed that when cotton cv DPL-51 (Upland cotton) was grown in controlled environments with natural solar radiation, flower and fruit retention was very low at an ambient temperature from 31.3 to 33°C plus 5 or 7°C. It was concluded that cotton will be severely damaged by temperatures above those presently observed during midsummer in the cotton belt in USA. Also, they concluded that the grower could minimize boll abscission where high temperature and low humidity occur by growing heat-tolerant cultivars, proper planting date, adequate fertilization, optimum plant density, and applying suitable irrigation regimes which would avoid drought stress. Oosterhuis (1997) studied the reasons for low and variable cotton yields in Arkansas, with unusually high insect pressures and the development of the boll load during an exceptionally hot and dry August. Solutions to the problems are suggested i.e. selection of tolerant cultivars, effective and timely insect and weed control, adequate irrigation regime, use of proper crop monitoring techniques and application of plant growth regulators.

3.2. Response of Flower and Boll Development to Climate Factors before and after Anthesis Day

The effects of specific climatic factors during both pre- and post-anthesis periods on boll production and retention are mostly unknown. However, by determining the relationship of climatic factors with flower and boll production and retention, the overall level of production can be possibly predicted. Thus, an understanding of these relationships may help physiologists to determine control mechanisms of production in cotton plants (Sawan et al. 2005).

Daily records of the climatic factors (independent variables), were taken for each day during production stage in any season including two additional periods of 15 days before and after the production stage (Sawan et al. 2005).

In each season, the data of the dependent and independent variables (68 and 62 days) were regarded as the original file (a file which contains the daily recorded data for any variable during a specific period). Fifteen other files before and another 15 after the production stage were obtained by fixing the dependent variable data, while moving the independent variable data at steps each of 1 day (either before or after production stage) in a matter similar to a sliding role (Sawan et al. 2005). The following is an example (in the first season):

File	Data of any dependent variable (for each flowers and bolls)		Any independent variable (for each climatic factors)			
	Production stage		In case of original file and files before production stage		In case of original file and files after production stage	
	Date	Days	Date	Days	Date	Days
Original file	23 Jun-29Aug	68	23 Jun-29 Aug	68	23 Jun-29Aug	68
1st new file	23 Jun-29Aug	68	22 Jun-28 Aug	68	24 Jun-30Aug	68
2nd new file	23 Jun-29Aug	68	21 Jun-27 Aug	68	25 Jun-31Aug	68
15th newfile	23 Jun-29Aug	68	8 Jun-14 Aug	68	8 Jul -13 Sept	68

Thus, the climate data were organized into records according to the complete production stage (68 days the first year and 62 days the second year) and 15 day, 14 day, 13 day,.and 1 day periods both before and after the production stage. This produced 31 climate periods per year that were analyzed for their relationships with cotton flowering and boll production (Sawan et al. 2005).

3.2.1. Correlation Estimates

A. Results of the correlation between climatic factors and flower and boll production during the 15 day periods before flowering day (Tables 4 and 5) revealed the following (Sawan et al. 2005):

First season. Daily evaporation and sunshine duration showed consistent negative and statistically significant correlations with both flower and boll production for each of the 15 moving window periods before anthesis (Table 4). Evaporation appeared to be the most important climate factor affecting flower and boll production.

Daily maximum and minimum humidity showed consistent positive and statistically significant correlations with both flower and boll production in most of the 15 moving window periods before anthesis (Table 4) (Sawan et al. 2005). Maximum daily temperature showed low but significant negative correlation with flower production during the 2-5, 8, and 10 day periods before anthesis. Minimum daily temperatures generally showed insignificant correlation with both production variables. The diurnal temperature range showed few correlations with flower and boll production. Daily soil surface temperature at 0600 h showed a significant positive correlation with boll production during the period extending from the 11-15 day period before anthesis, while its effect on flowering was confined only to the 12 and the 15 day periods prior to anthesis. Daily soil surface temperature at 1800 h showed a significant negative correlation with flower production during the 2-10 day periods before anthesis (Sawan et al. 2005).

Second season. Daily Evaporation, the diurnal temperature range, and sunshine duration were negatively and significantly correlated with both flower and boll production in all the 15 day periods, while maximum daily temperature was negatively and significantly related to flower and boll formation during the 2- 5 day periods before anthesis (Table 5) (Sawan et al. 2005). Minimum daily temperature showed positive and statistically significant correlations with both production variables only during the 9-15 day periods before anthesis, while daily minimum humidity showed the same correlation trend in all the 15 moving window periods before anthesis. Daily soil surface temperature at 0600 h was positively and significantly correlated with flower and boll production for the 12, 14, and 15 day periods prior to anthesis only. Daily soil surface temperature at 1800 h showed negative and significant correlations with both production variables only during the first and second day periods before flowering. Daily maximum humidity showed insignificant correlation with both flower and boll production except for one day period only (the 15th day) (Sawan et al. 2005).

Table 4. Simple correlation coefficients (r) between climatic factors and number of flower and harvested bolls in initial time (0) and each of the 15–day periods before flowering in the first season (I)

Climate period		Air temp. (°C)			Evap. (mm d^{-1})	Surface soil temp. (°C)		Sunshine duration (h d^{-1})	Humidity (%)	
		Max. (X_1)	Min. (X_2)	Max-Min$^\bullet$ (X_3)	(X_4)	0600 h (X_5)	1800 h (X_6)	(X_7)	Max. (X_8)	Min. (X_9)
0$^\#$	Flower	-0.07	-0.06	-0.03	-0.56**	-0.01	-0.20	-0.25*	0.40**	0.14
	Boll	-0.03	-0.07	-0.01	-0.53**	-0.06	-0.16	-0.14	0.37**	0.10
1	Flower	-0.15	-0.08	-0.11	-0.64**	-0.01	-0.17	-0.30*	0.39**	0.20
	Boll	-0.07	-0.08	-0.02	-0.58**	-0.06	-0.10	-0.23*	0.36**	0.13
2	Flower	-0.26*	-0.10	-0.22	-0.69**	-0.07	-0.30*	-0.35*	0.42**	0.30*
	Boll	-0.18	-0.08	-0.14	-0.64**	-0.05	-0.21	-0.25*	0.40**	0.20
3	Flower	-0.28*	-0.02	-0.31**	-0.72**	0.15	-0.29*	-0.37**	0.46**	0.35**
	Boll	-0.19	-0.02	-0.21	-0.65**	0.11	-0.20	-0.30*	0.37**	0.25*
4	Flower	-0.26*	-0.03	-0.26*	-0.67**	0.08	-0.24*	-0.41**	0.46**	0.35**
	Boll	-0.21	-0.04	-0.21	-0.63**	0.04	-0.18	-0.35**	0.39**	0.29*
5	Flower	-0.27*	-0.02	-0.27*	-0.68**	0.16	-0.29*	-0.45**	0.49**	0.38**
	Boll	-0.22	0.00	-0.24*	-0.63**	0.16	-0.21	-0.39**	0.44**	0.32**
6	Flower	-0.21	0.05	-0.25*	-0.73**	0.16	-0.28*	-0.46**	0.47**	0.42**
	Boll	-0.15	0.08	-0.21	-0.67**	0.19	-0.19	-0.46**	0.43**	0.35**
7	Flower	-0.17	-0.01	-0.17	-0.69**	0.10	-0.27*	-0.43**	0.46**	0.35**
	Boll	-0.11	-0.06	-0.15	-0.64**	0.14	-0.19	-0.46**	0.43**	0.32**
8	Flower	-0.24*	-0.03	-0.24*	-0.71**	0.09	-0.30*	-0.44**	0.45**	0.45**
	Boll	-0.14	0.04	-0.17	-0.63**	0.16	-0.17	-0.48**	0.44**	0.39**
9	Flower	-0.23	-0.10	-0.19	-0.68**	0.05	-0.33**	-0.32**	0.43**	0.44**
	Boll	-0.14	0.04	-0.17	-0.61**	0.15	-0.21	-0.40**	0.42**	0.41**
10	Flower	-0.26*	0.05	-0.30*	-0.67**	0.13	-0.29*	-0.29*	0.40**	0.48**
	Boll	-0.14	0.13	-0.22	-0.58**	0.22	-0.17	-0.36**	0.46**	0.41**
11	Flower	-0.20	0.10	-0.27*	-0.62**	0.21	-0.19	-0.29*	0.42**	0.44**
	Boll	-0.04	0.22	-0.16	-0.53**	0.27*	-0.04	-0.38**	0.45**	0.36**
12	Flower	-0.17	0.16	-0.26*	-0.62**	0.29*	-0.15	-0.40**	0.44**	0.45**
	Boll	0.00	0.25*	-0.13	-0.51**	0.35**	-0.04	-0.45**	0.40**	0.30*
13	Flower	-0.13	0.16	-0.22	-0.62**	0.23	-0.12	-0.42**	0.43**	0.45**
	Boll	0.00	0.22	-0.11	-0.51**	0.30*	-0.03	-0.49**	0.41**	0.33**
14	Flower	-0.08	0.18	-0.18	-0.56**	0.21	-0.15	-0.44**	0.41**	0.46**
	Boll	0.01	0.21	-0.10	-0.47**	0.26*	-0.09	-0.49**	0.42**	0.33**
15	Flower	-0.08	0.22	-0.21	-0.51**	0.24*	-0.22	-0.42**	0.39**	0.38**
	Boll	-0.03	0.19	-0.13	-0.45**	0.24*	-0.17	-0.44**	0.43**	0.30*

*: Significant at 5% level and **: significant at 1% level.

$^\#$ 0 = Initial time.

$^\bullet$ diurnal temperature range.

(Sawan et al., 2005)

Response of Flower and Boll Development to Climatic Factors ... 19

Table 5. Simple correlation coefficients (r) between climatic factors[z] and number of flower and harvested bolls in initial time (0) and each of the 15–day periods before flowering in the second season (II)

Climate period		Air temp. (°C)			Evap. (mm d⁻¹)	Surface soil temp. (°C)		Sunshine duration (h d⁻¹)	Humidity (%)	
		Max. (X_1)	Min. (X_2)	Max-Min[•] (X_3)	(X_4)	0600 h (X_5)	1800 h (X_6)	(X_7)	Max. (X_8)	Min. (X_9)
0[#]	Flower	-0.42**	0.00	-0.36**	-0.61**	-0.14	-0.37**	-0.37**	0.01	0.45**
	Boll	-0.42**	0.02	-0.37**	-0.59**	-0.13	-0.36**	-0.36**	0.01	0.46**
1	Flower	-0.42**	0.10	-0.42**	-0.63**	-0.08	-0.29*	-0.41**	0.05	0.48**
	Boll	-0.41**	0.11	-0.42**	-0.62**	-0.07	-0.28*	-0.41**	0.05	0.47**
2	Flower	-0.40**	0.08	-0.43**	-0.65**	-0.09	-0.27*	-0.39**	0.02	0.49**
	Boll	-0.40**	0.08	-0.43**	-0.64**	-0.08	-0.26*	-0.40**	0.03	0.49**
3	Flower	-0.38**	0.13	-0.43**	-0.61**	-0.06	-0.17	-0.38**	0.00	0.45**
	Boll	-0.37**	0.15	-0.44**	-0.61**	-0.05	-0.15	-0.38**	0.01	0.46**
4	Flower	-0.36**	0.17	-0.41**	-0.61**	-0.04	-0.18	-0.38**	0.02	0.45**
	Boll	-0.35**	0.18	-0.41**	-0.60**	-0.03	-0.16	-0.36**	0.03	0.44**
5	Flower	-0.30*	0.13	-0.36**	-0.60**	-0.07	-0.23	-0.32**	-0.05	0.43**
	Boll	-0.28*	0.15	-0.35**	-0.58**	-0.05	-0.21	-0.31**	-0.05	0.41**
6	Flower	-0.24	0.21	-0.38**	-0.61**	-0.02	-0.12	-0.28*	0.02	0.40**
	Boll	-0.22	0.24	-0.38**	-0.59**	0.00	-0.07	-0.29*	0.02	0.40**
7	Flower	-0.19	0.23	-0.29*	-0.54**	-0.03	-0.05	-0.26*	-0.04	0.32**
	Boll	-0.18	0.23	-0.27*	-0.53**	-0.02	-0.03	-0.27*	-0.04	0.30*
8	Flower	-0.15	0.24	-0.25*	-0.52**	-0.03	-0.07	-0.24*	-0.05	0.28*
	Boll	-0.14	0.22	-0.22	-0.51**	-0.03	-0.06	-0.22*	-0.05	0.26*
9	Flower	-0.16	0.34**	-0.32**	-0.56**	0.08	-0.02	-0.25*	0.05	0.30*
	Boll	-0.14	0.34**	-0.31**	-0.56**	0.09	-0.01	-0.23*	0.07	0.29*
10	Flower	-0.16	0.31**	-0.30*	-0.56**	0.11	-0.06	-0.27*	0.11	0.33**
	Boll	-0.14	0.28*	-0.27*	-0.55**	0.09	-0.07	-0.25*	0.09	0.31**
11	Flower	-0.16	0.31**	-0.27*	-0.55**	0.10	-0.02	-0.31**	0.08	0.32**
	Boll	-0.15	0.29*	-0.26*	-0.53**	0.10	0.00	-0.29*	0.08	0.29*
12	Flower	-0.17	0.44**	-0.37**	-0.57**	0.26*	0.02	-0.36**	0.17	0.34**
	Boll	-0.17	0.42**	-0.36**	-0.55**	0.25*	0.01	-0.34**	0.16	0.32**
13	Flower	-0.14	0.40**	-0.33**	-0.56**	0.21	0.03	-0.28*	0.10	0.34**
	Boll	-0.15	0.38**	-0.34**	-0.56**	0.21	0.01	-0.27*	0.09	0.33**
14	Flower	-0.19	0.39**	-0.38**	-0.59**	0.25*	0.04	-0.34**	0.16	0.35**
	Boll	-0.20	0.39**	-0.40**	-0.59**	0.26*	0.03	-0.36**	0.17	0.36**
15	Flower	-0.24	0.49**	-0.45**	-0.62**	0.37**	0.16	-0.38**	0.27*	0.42**
	Boll	-0.24	0.51**	-0.48**	-0.63**	0.40**	0.15	-0.40**	0.26*	0.43**

*: Significant at 5% level and **: significant at 1% level.
[#] 0 = Initial time.
[•] diurnal temperature range.
[z] Wind speed did not show significant effect upon the studied production variables, so it is not reported.
(Sawan et al., 2005)

Generally, the results in the two seasons indicated that daily evaporation, sunshine duration and minimum humidity were the most effective and consistent climatic factors, which exhibited significant relationships with the production variables for all the 15 day periods before anthesis in both seasons (Sawan et al., 2005).

The factors in this study which had been found to be associated with boll development are the climatic factors that would influence water loss between plant and atmosphere (low evaporation demand, high humidity, and shorter solar duration). This can lead to direct effects on the fruiting forms themselves and inhibitory effects on mid-afternoon photosynthetic rates even under well-watered conditions. Boyer et al. (1980) found that soybean plants with ample water supplies can experience water deficits due to high transpiration rates. Also, Human et al. (1990) stated that, when sunflower plants were grown under controlled temperature regimes, water stress during budding, anthesis and seed filling, the CO_2 uptake rate per unit leaf area as well as total uptake rate per plant, significantly diminished with stress, while this effect resulted in a significant decrease in yield per plant.

B. The correlation between climatic factors and each of boll production and boll retention over a period of 15 day periods after flowering (boll setting) day (Tables 6 and 7) (Sawan et al. 2005) revealed the following:

First season. Daily evaporation showed significant negative correlation with number of bolls for all the 15 day periods after flowering (Table 6) (Sawan et al. 2005). Meanwhile its relationship with retention ratio was positive and significant in the 9-15 day periods after flowering. Daily sunshine duration was positively and significantly correlated with boll retention ratio during the 5-13 day periods after flowering. Daily maximum humidity had a significant positive correlation with the number of bolls during the first 8 day periods after flowering, while daily minimum humidity had the same correlation for only the 11, and 12 day periods after flowering. Daily maximum and minimum temperatures and the diurnal temperature range, as well as soil surface temperature at 1800 did not show significant relationships with both number of bolls and retention ratio. Daily soil surface temperature at 0600 h had a significant negative correlation with boll retention ratio during the 3-7 day periods after anthesis.

Second season. Daily evaporation, soil surface temperature at 1800 h, and sunshine duration had a significant negative correlation with number of bolls in all the 15 day periods after anthesis (Table 7) (Sawan et al. 2005). Daily maximum and minimum temperatures and the diurnal temperature range, and soil surface temperature at 0600 h had a negative correlation with boll production. Their significant effects were observed during the 1, and 10-15 day periods for maximum temperature, and the 1-5, and 9-12 day periods for the diurnal temperatures range. Meanwhile, the daily minimum temperature and soil surface temperature at 0600 h had a significant negative correlation only during the 13-15 day periods. Daily minimum humidity had a significant positive correlation with number of bolls during the first 5 day periods, and the 9-15 day periods after anthesis. Daily maximum humidity showed no significant relation to number of bolls produced, and further no significant relation was observed between any of the studied climatic factors and boll retention ratio (Sawan et al. 2005).

The results in the two seasons indicated that evaporation and humidity, followed by sunshine duration had obvious correlation with boll production (Sawan et al. 2005). From the results obtained, it appeared that the effects of air temperature, and soil surface temperature tended to be masked in the first season, i.e. did not show any significant effects in the first season on the number of bolls per plant. However, these effects were found to be significant in the second season. These seasonal differences in the impacts of the previously mentioned climatic factors on the number of bolls per plant are most likely ascribed to the sensible variation in evaporation values in the two studied seasons where their means were 10.2 mm.d^{-1} and 5.9 mm d^{-1} in the first and second seasons, respectively (Sawan et al. 2005).

There is an important question concerning, forecasting when evaporation values would mask the effect of the previous climatic factors (Sawan et al. 2005). The answer would be possibly achieved through relating humidity values to evaporation values which are naturally liable to some fluctuations from one season to another (Sawan et al. 2005). It was found that the ratio between the mean of maximum humidity and the mean of evaporation in the first season was $85.8/10.2 = 8.37$, while in the second season this ratio was 12.4. On the other hand, the ratio between the mean minimum humidity and the mean of evaporation in the first season was $30.8/10.2 = 3.02$, while in the second season this ratio was 6.75 (Table 6) (Sawan et al. 2005). From these ratios it seems that minimum humidity which is closely related to evaporation is more sensitive than the ratio between maximum humidity and evaporation. It can be seen from the results and formulas that when the ratio between minimum humidity and evaporation is small (3:1), the effects of air temperature, and soil surface temperature were hindered by the effect of evaporation, i.e. the effect of these climatic factors were not significant. However, when this ratio is high (6:1), the effects of these factors were found to be significant. Accordingly, it could be generally stated that the effects of air, and soil surface temperatures could be masked by evaporation when the ratio between minimum humidity and evaporation is less than 4:1 (Sawan et al. 2005).

Table 6. Simple correlation coefficient (r) values between climatic factors and number of harvested bolls and retention ratio in initial time (0) and each of the 15–day periods after flowering in the first season (I)

Climate period		Air temp. (°C)			Evap. (mm d^{-1})	Surface soil temp. (°C)		Sunshine duration (h d^{-1})	Humidity (%)	
		Max. (X_1)	Min. (X_2)	Max.-Min$^{\bullet}$. (X_3)	(X_4)	0600 h (X_5)	1800 h (X_6)	(X_7)	Max. (X_8)	Min. (X_9)
0[#]	Retention ratio•	-0.05	-0.03	-0.03	-0.10	-0.11	0.10	0.20	-0.04	-0.02
	No. of bolls	-0.03	-0.07	-0.01	-0.53**	-0.06	-0.16	-0.14	0.37**	0.10
1	Retention ratio	-0.07	-0.08	-0.01	-0.10	-0.16	0.04	0.15	0.04	0.05
	No. of bolls	0.02	-0.08	0.08	-0.49**	-0.09	-0.05	-0.20	0.35**	0.09
2	Retention ratio	-0.08	-0.14	0.02	-0.08	-0.19	0.03	0.17	0.02	-0.02
	No. of bolls	0.02	-0.04	0.07	-0.46**	-0.06	-0.01	-0.19	0.33**	0.09
3	Retention ratio	-0.09	-0.21	0.06	-0.08	-0.24*	0.02	0.19	0.01	-0.10
	No. of bolls	0.03	-0.03	0.06	-0.44**	-0.04	0.05	-0.18	0.32**	0.08
4	Retention ratio	-0.05	-0.20	0.09	-0.01	-0.24*	0.01	0.22	0.00	-0.15
	No. of bolls	0.01	-0.05	0.05	-0.40**	-0.03	0.04	-0.16	0.31*	0.08
5	Retention ratio	-0.03	-0.21	0.13	0.07	-0.25*	0.00	0.26*	-0.02	-0.22
	No. of bolls	0.00	-0.07	0.05	-0.37**	-0.02	0.03	-0.13	0.29*	0.07
6	Retention ratio	0.01	-0.19	0.15	0.12	-0.24*	0.02	0.27*	-0.03	-0.20
	No. of bolls	-0.01	-0.08	0.04	-0.38**	-0.02	0.04	-0.15	0.31*	0.13
7	Retention ratio	0.05	-0.17	0.17	0.18	-0.25*	0.05	0.29*	-0.02	-0.21
	No. of bolls	-0.03	-0.09	0.03	-0.39**	-0.04	0.06	-0.14	0.34**	0.18
8	Retention ratio	0.06	-0.08	0.13	0.21	-0.20	0.07	0.28*	-0.06	-0.19
	No. of bolls	-0.05	-0.07	-0.01	-0.35**	-0.02	0.02	-0.17	0.28*	0.17
9	Retention ratio	0.08	0.00	0.08	0.26*	-0.14	0.08	0.29*	-0.12	-0.20
	No. of bolls	-0.08	-0.06	-0.05	-0.33**	-0.01	0.00	-0.23	0.20	0.16
10	Retention ratio	0.06	-0.02	0.05	0.27*	-0.13	0.09	0.27*	-0.10	-0.08
	No. of bolls	-0.11	-0.10	-0.07	-0.34**	-0.03	-0.03	-0.19	0.18	0.21
11	Retention ratio	0.04	-0.04	0.08	0.28*	-0.12	0.08	0.26*	-0.09	-0.05
	No. of bolls	-0.18	-0.18	-0.06	-0.37**	-0.10	-0.04	-0.14	0.15	0.28*
12	Retention ratio	0.02	0.01	-0.08	0.32**	-0.05	0.05	0.25*	-0.08	-0.03
	No. of bolls	-0.17	-0.13	-0.08	-0.32**	-0.06	-0.07	-0.11	0.16	0.24*
13	Retention ratio	-0.04	0.04	-0.09	0.38**	0.00	0.01	0.27*	-0.09	-0.02
	No. of bolls	-0.15	-0.09	-0.09	-0.29*	-0.03	-0.10	-0.08	0.18	0.20
14	Retention ratio	-0.07	0.04	-0.13	0.34**	0.06	-0.02	0.18	-0.08	-0.01
	No. of bolls	-0.15	-0.10	-0.10	-0.28*	-0.01	-0.10	-0.15	0.17	0.17
15	Retention ratio	-0.13	0.03	-0.18	0.33**	0.09	-0.04	0.06	-0.07	0.00
	No. of bolls	-0.16	-0.10	-0.11	-0.28*	0.00	-0.11	-0.13	0.17	0.15

* and ** Significant at 5% and 1% levels of significance, respectively.

[#] 0 = Initial time

• Retention ratio: (the number of retained bolls obtained from the total number of each daily tagged flowers in all selected plants at harvest/each daily number of tagged flowers in all selected plants) x 100.

♦ diurnal temperature range.

(Sawan et al., 2005)

Table 7. Simple correlation coefficient (r) values between climatic factors[z] and number of harvested bolls and retention ratio in initial time (0) and each of the 15–day periods after flowering in the second season (II)

Climate period		Air temp. (°C)			Evap. (mm d^{-1})	Surface soil temp. (°C)		Sunshine duration (h d^{-1})	Humidity (%)	
		Max. (X_1)	Min. (X_2)	Max.-Min.[•] (X_3)	(X_4)	0600 h (X_5)	1800 h (X_6)	(X_7)	Max. (X_8)	Min. (X_9)
0[#]	Retention ratio[•]	-0.04	0.20	-0.31*	-0.14	0.12	-0.20	0.01	-0.04	0.17
	No. of bolls	-0.42**	0.02	-0.37**	-0.59**	-0.13	-0.36**	-0.36**	0.01	0.46**
1	Retention ratio	-0.10	-0.03	-0.22	-0.21	-0.15	-0.05	-0.04	-0.02	0.23
	No. of bolls	-0.25*	-0.01	-0.36**	-0.63**	-0.15	-0.30*	-0.25*	0.06	0.44**
2	Retention ratio	-0.15	-0.06	-0.10	-0.15	-0.08	-0.21	-0.01	-0.04	0.12
	No. of bolls	-0.18	-0.01	-0.34**	-0.65**	-0.11	-0.25*	-0.32*	0.13	0.43**
3	Retention ratio	-0.03	-0.01	-0.02	-0.21	-0.01	-0.17	-0.08	0.09	0.12
	No. of bolls	-0.15	-0.06	-0.30*	-0.62**	-0.05	-0.28*	-0.31*	0.14	0.33**
4	Retention ratio	0.08	-0.02	0.07	-0.09	-0.03	-0.09	-0.10	0.05	-0.04
	No. of bolls	-0.15	-0.05	-0.28*	-0.63**	-0.06	-0.25*	-0.33**	0.15	0.32*
5	Retention ratio	0.23	-0.03	0.12	-0.06	-0.06	-0.01	-0.11	0.01	-0.16
	No. of bolls	-0.14	-0.05	-0.25*	-0.62**	-0.06	-0.24*	-0.35**	0.15	0.31*
6	Retention ratio	0.09	-0.08	0.12	-0.09	-0.07	-0.01	-0.09	0.00	-0.05
	No. of bolls	-0.15	-0.04	-0.22	-0.61**	-0.08	-0.25*	-0.34**	0.13	0.22
7	Retention ratio	-0.03	-0.12	0.12	-0.10	-0.11	-0.01	-0.04	-0.03	0.02
	No. of bolls	-0.15	-0.02	-0.19	-0.60**	-0.10	-0.29*	-0.32*	0.10	0.18
8	Retention ratio	-0.02	0.05	0.03	-0.10	-0.04	-0.03	-0.02	-0.01	0.01
	No. of bolls	-0.20	-0.03	-0.23	-0.61**	-0.10	-0.28*	-0.32*	0.19	0.22
9	Retention ratio	-0.02	0.13	-0.05	-0.10	0.08	-0.05	-0.01	0.03	0.00
	No. of bolls	-0.24	-0.04	-0.29*	-0.62**	-0.11	-0.30*	-0.33**	0.13	0.27*
10	Retention ratio	-0.04	0.12	-0.08	-0.09	0.05	0.11	-0.02	0.04	0.02
	No. of bolls	-0.27*	-0.07	-0.30*	-0.60**	-0.16	-0.34**	-0.34**	0.11	0.26*
11	Retention ratio	-0.07	0.10	-0.10	-0.08	0.03	0.20	-0.03	0.05	0.04
	No. of bolls	-0.30*	-0.12	-0.30*	-0.61**	-0.18	-0.39**	-0.36**	0.10	0.27*
12	Retention ratio	-0.11	0.09	-0.14	-0.11	0.04	0.13	-0.08	0.11	0.09
	No. of bolls	-0.32*	-0.19	-0.26*	-0.60**	-0.22	-0.42**	-0.37**	0.09	0.27*
13	Retention ratio	-0.14	0.09	-0.17	-0.18	0.06	-0.06	-0.14	0.16	0.12
	No. of bolls	-0.33**	-0.26*	-0.23	-0.59**	-0.28*	-0.48**	-0.39**	0.08	0.27*
14	Retention ratio	-0.11	-0.04	-0.10	-0.13	-0.15	-0.05	-0.09	0.01	0.12
	No. of bolls	-0.34**	-0.32*	-0.21	-0.61**	-0.32*	-0.48**	-0.38**	0.06	0.27*
15	Retention ratio	-0.08	-0.11	0.02	-0.08	-0.22	-0.05	-0.02	-0.03	0.12
	No. of bolls	-0.35*	-0.37**	-0.18	-0.61**	-0.38**	-0.48**	-0.37**	0.03	0.27*

* and ** Significant at 5% and 1% levels of significance, respectively.

[#] 0 = Initial time

• Retention ratio: (the number of retained bolls obtained from the total number of each daily tagged flowers in all selected plants at harvest/each daily number of tagged flowers in all selected plants) x 100.

♦ diurnal temperature range.

[z] Wind speed did not show significant effect upon the studied production variables, so it is not reported.

(Sawan et al., 2005)

Evaporation appeared to be the most important climatic factor (in each of the 15-day periods both prior to and after initiation of individual bolls) affecting number of flowers or harvested bolls in Egyptian cotton (Sawan et al. 2005). High daily evaporation rates could result in water stress that would slow growth and increase shedding rate of flowers and bolls (Sawan et al. 2005). The second most important climatic factor in our study was humidity. Effect of maximum humidity varied markedly from the first season to the second one, where it was significantly correlated with the dependent variables in the first season, while the inverse pattern was true in the second season (Sawan et al. 2005). This diverse effect may be due to the differences in the values of this factor in the two seasons; where it was on average 87% in the first season, and only 73% in the second season (Table 1) (Sawan et al. 2005). Also when the average value of minimum humidity exceeded half the average value of maximum humidity, the minimum humidity can substitute the maximum humidity on affecting number of flowers or harvested bolls. In the first season (Table 1) (Sawan et al. 2005) the average value of minimum humidity was less than half of the value of maximum humidity (30.2/85.6 = 0.35), while in the second season it was higher than half of maximum humidity (39.1/72.9 = 0.54) (Sawan et al. 2005).

The third most important climatic factor in our study was sunshine duration, which showed a significant negative relationship with boll production (Sawan et al., 2005). The r values of (Tables 4-7) (Sawan et al. 2005) indicated that the relationship between the dependent and independent variables preceding flowering (production stage) generally exceeded in value the relationship between them during the entire and late periods of production stage. In fact, understanding the effects of climatic factors on cotton production during the previously mentioned periods would have marked consequences on the overall level of cotton production, which could be predictable depending on those relationships (Sawan et al., 2005).

3.2.2. Regression Models

An attempt was made to investigate the effect of climatic factors on cotton production via prediction equations including the important climatic factors responsible for the majority of total variability in cotton flower and boll production (Sawan et al. 2005). Hence, regression models were established using the stepwise multiple regression technique to express the relationship between each of the number of flowers and bolls/plant and boll retention ratio (Y), with the climatic factors, for each of the a) 5, b) 10, and c) 15 day periods either prior to or after initiation of individual bolls (Tables 8 and 9) (Sawan et al. 2005).

Concerning the effect of prior days the results indicated that evaporation, sunshine duration, and the diurnal temperature range were the most effective and consistent climatic factors affecting cotton flower and boll production (Table 8) (Sawan et al. 2005). The fourth effective climatic factor in this respect was minimum humidity. On the other hand, for the periods after flower the results obtained from the equations (Table 9) (Sawan et al. 2005) indicated that evaporation was the most effective and consistent climatic factor affecting number of harvested bolls.

Regression models demonstrate each independent variable under study as an efficient and important factor (Sawan et al. 2005). Meanwhile, they explained a sensible proportion of the variation in flower and boll production, as indicated by their R^2, which ranged between 0.14-0.62, where most of R^2 prior to flower opening were about 0.50 and after flowering all but one are less than 0.50 (Sawan et al. 2005).These results agree with Miller et al. (1996) in their regression study of the relation of yield with rainfall and temperature. They suggested that the other 0.50 of variation related to management practices, which can be the same in this study. Also, the regression models indicated that the relationships between the number of flowers and bolls per plant and the studied climatic factors for the 15 day period before or after flowering (Y_3) in each season explained the highly significant magnitude of variation $(P < 0.05)$. The R^2 values for the 15 day periods before and after flowering were higher than most of those obtained for each of the 5 and the 10 day periods before or after flowering. This clarifies that the effects of the climatic factors during the 15 day periods before or after flowering are very important for Egyptian cotton boll production and retention. Thus, an accurate climatic forecast for the effect of these 15 day periods provides an opportunity to avoid any possible adverse effects of unusual climatic conditions before flowering or after boll formation by utilizing additional treatments and/or adopting proper precautions to avoid flower and boll reduction.

The main climatic factors from this study (Sawan et al. 2005) affecting the number of flowers and bolls, and by implication yield, is evaporation, sunshine duration and minimum humidity, with evaporation (water stress) being by far the most important factor (Sawan et al. 2005). Various activities have been suggested to partially overcome water stress. Temperature conditions during the reproduction growth stage of cotton in Egypt do not appear to limit growth even though they are above the optimum for cotton growth (Sawan et al. 2005). This is contradictory to the finding of Holaday et al. (1997). A possible reason for that contradiction is that the effects of evaporation rate and humidity were not taken into consideration in the research studies conducted by other researchers in other countries.

Table 8. The models obtained for the number of flowers and bolls per plant as functions of the climatic data derived from the 5, 10, and 15 day periods prior to flower opening in the two seasons (I, II)

Season	Model [z]	R^2	Significance
First			
Flower			
	$Y_1 = 55.75 + 0.86X_3 - 2.09X_4 - 2.23X_7$	0.51	**
	$Y_2 = 26.76 - 5.45X_4 + 1.76X_9$	0.42	**
	$Y_3 = 43.37 - 1.02X_4 - 2.61X_7 + 0.20X_8$	0.52	**
Boll			
	$Y_1 = 43.69 + 0.34X_3 - 1.71X_4 - 1.44X_7$	0.43	**
	$Y_2 = 40.11 - 1.82X_4 - 1.36X_7 + 0.10X_8$	0.48	**
	$Y_3 = 31.00 - 0.60X_4 - 2.62X_7 + 0.23X_8$	0.47	**
Second			
Flower			
	$Y_1 = 18.58 + 0.39X_3 - 0.22X_4 - 1.19X_7 + 0.17X_9$	0.54	**
	$Y_2 = 16.21 + 0.63X_3 - 0.20X_4 - 1.24X_7 + 0.16X_9$	0.61	**
	$Y_3 = 14.72 + 0.51X_3 - 0.20X_4 - 0.85X_7 + 0.17X_9$	0.58	**
Boll			
	$Y_1 = 25.83 + 0.50X_3 - 0.26X_4 - 1.95X_7 + 0.15X_9$	0.61	**
	$Y_2 = 19.65 + 0.62X_3 - 0.25X_4 - 1.44X_7 + 0.12X_9$	0.60	**
	$Y_3 = 15.83 + 0.60X_3 - 0.22X_4 - 1.26X_7 + 0.14X_9$	0.59	**

[z] Where Y_1, Y_2, Y_3 = number of flowers or bolls per plant at the 5, 10 and 15 day periods before flow respectively, X_2 = minimum temperature (°C), X_3 = diurnal temperature range (°C), X_4 = evaporation X_7 = sunshine duration (h day^{-1}), X_8 = maximum humidity (%) and X_9 = minimum humidity (%). (Sawan et al., 2005)

Table 9. The models obtained for the number of bolls per plant as functions of the climatic data derived from the 5, 10, and 15 day periods after flower opening in the two seasons (I, II)

Season	Model [z]	R^2	Significance
First	$Y_1 = 16.38 - 0.41X_4$	0.14	**
	$Y_2 = 16.43 - 0.41X_4$	0.14	**
	$Y_3 = 27.83 - 0.60X_4 - 0.88X_9$	0.15	**
Second	$Y_1 = 23.96 - 0.47X_4 - 0.77X_8$	0.44	**
	$Y_2 = 18.72 - 0.58X_4$	0.34	**
	$Y_3 = 56.09 - 2.51X_4 - 0.49X_6 - 1.67X_7$	0.56	**

[z] Where Y_1, Y_2, Y_3 = number of bolls per plant at the 5, 10, and 15 day periods after flowering, respectively, X_4 = evaporation (mm day^{-1}), X_6 = soil surface temperature (°C) at 1800, X_7 = sunshine duration (h day^{-1}), X_8 = maximum humidity (%) and X_9 = minimum humidity (%).
(Sawan et al., 2005)

The matter of fact is that temperature and evaporation are closely related to each other to such an extent that the higher evaporation rate could possibly mask the effect of temperature. Sunshine duration and minimum humidity appeared to have secondary effects, yet they are in fact important players (Sawan et al. 2005). The importance of sunshine duration has been alluded to by Moseley et al. (1994) and Oosterhuis (1997). Also, Mergeai and Demol (1991) found that cotton yield was assisted by intermediate relative humidity.

3.3. Effect of Climatic Factors during the Development Periods of Flowering and Boll Formation on the Production of Cotton

Observation used in the statistical analysis was obtained during the entire production stage of flowering and boll development (60 days for each season, 29 June to 27 August).

Independent variables, their range and mean values for the two seasons and during the periods of flower and boll production are listed in Table 1 (Sawan et al. 2005). Both flower number and boll production show the higher value in the third and fourth quarters of production stage, accounting for about 70% of total production during the first season and about 80% of the total in the second season (Sawan et al. 1999).

Linear correlation between the climatic factors and the studied characteristics, i.e. flower, boll production and boll retention ratio, were calculated based on quarters of the production stage for each season. Significant relationships ($\leq$ 0.15) are shown in Tables 2 and 3 (Sawan et al. 1999). Examining these tables, it is clear that the fourth quarter of production stage consistently exhibited the highest R^2 values regardless of the second quarter for boll retention ratio; however, less data pairs were used (n = 30 for combined data of the fourth quarter "n = 15 for each quarter of each season") to calculate the relations.

Results obtained from the four quarters of the production period for each season separately and for the combined data of the two seasons, indicated that relationships varied markedly from one season to another. This may be due to the differences between the climatic factors in the two seasons; as illustrated by its ranges and means shown in Table 1 (Sawan et al. 2005). For example, maximum temperature and surface soil temperature at 1800 h did not show significant effects in the first season, while this trend differed in the second season.

Multiple linear regression equations obtained from data of the fourth quarter, for:

1. Flower production,
$Y = 160.0 + 11.28X_1 - 4.45X_3 - 2.93X_4 - 5.05X_5 - 11.3X_6 - 0.962X_8 + 2.36X9$
And $R^2 = 0.672**$

2. Boll production,
$Y = 125.4 + 13.74X_1 - 6.76X_3 - 4.34X_4 - 6.59X_5 - 10.3X_6 - 1.25X_8 + 2.16X9$
With an $R^2 = 0.747**$

3. Boll retention ratio,
$Y = 81.93 - 0.272X_3 - 2.98X_4 + 3.80X_7 - 0.210X_8 - 0.153X_9$
And its $R^2 = 0.615**$
The equation obtained from data of the second quarter of production stage for boll retention ratio,
$Y = 92.81 - 0.107X_3 - 0.453X_4 + o.298X_7 - 0.194X_8 + 0.239X_9$
And $R^2 = 0.737**$

R^2 values for these equations ranged from 0.615 to 0.747 (Sawan et al. 1999). It could be concluded that these equations may predict flower and boll

Response of Flower and Boll Development to Climatic Factors ...   29

production and boll retention ratio from the fourth quarter period within about 62 to 75% of its actual means. Therefore, these equations seem to have practical value. R^2 between the fourth quarter and the entire production period of the two seasons for each of flower, boll production, and boll retention ratio were large (0.266, 0.325, and 0.279 respectively). These differences are sufficiently large to make a wide gap under a typical field sampling situation. This could be due to the high percentage of flower and boll production for the fourth quarter.

Table 10. Significant simple correlation values between the climatic factors and flower, boll production and boll retention ratio due to quarters of production stage

Climatic factors		Flower				Boll				Ratio:Bolls/Flowers (100)			
		1st	2nd	3rd	4th	1st	2nd	3rd	4th	1st	2nd	3rd	4th
First season (n by quarter = 15)													
MaxTemp °C,	(X_1)	n.s.	n.s.	n.s.	n.s.	n.s.	n.s.	n.s.	n.s.	n.s.	n.s.	n.s.	n.s..
Min Temp °C,	(X_2)	0.516*	0.607*	n.s.	n.s.	0.561*	0.638**	n.s.	n.s.	n.s.	0.680**	n.s.	n.s.
Max-Min °C,	(X_3)	n.s.	n.s.	0.538*	n.s.	n.s.	n.s.	0.494*	n.s.	0.515*	n.s.	n.s.	n.s.
Evapor. mm/d,	(X_4)	0.512*	0-.598*	n.s.	0.424++	0.397+	-0.500*	-.0321+	n.s.	n.s.	-0.387+	-0.287+	n.s.
0600 h Temp. °C,	(X_5)	-0.352+	0.534*	-0.358+	0.301+	0.402+	0.516*	-0.441++	n.s.	n.s.	0.440++	n.s.	-.292+
1800 h Temp. °C,	(X_6)	n.s.	n.s.	n.s.	n.s.	n.s.	n.s.	n.s.	n.s.	n.s.	n.s.	n.s.	n.s.
Sunshine h/d,	(X_7)	n.s.	n.s.	0.346*	n.s.	n.s.	n.s.	n.s.	0.430++	n.s.	n.s.	n.s.	0.480*
Max Hum %,	(X_8)	-0.316*	-0.260*	0.461++	0.283+	n.s.	n.s.	0.410++	n.s.	.389+	n.s.	n.s.	-0.322*
Min Hum %,	(X_9)	n.s.	0.309+	-0.436++	n.s.	n.s.	0.436++	-0.316++	n.s.	-0.473++	0.527*	n.s.	n.s.
Second season (n by quarter = 15)													
MaxTemp °C,	(X_1)	n.s.	n.s.	n.s.	-0.730**	n.s.	n.s.	n.s.	-0.654**	n.s.	n.s.	0.407++	n.s.
Min Temp °C,	(X_2)	n.s.	n.s.	n.s.	-0.451++	n.s.	n.s.	n.s.	-0.343+	n.s.	n.s.	n.s.	n.s.
Max-Min °C,	(X_3)	n.s.	n.s.	0.598*	n.s.	n.s.	n.s.	0.536*	n.s.	0.456++	-0.416+	n.s.	n.s.
Evapor. mm/d,	(X_4)	n.s.	n.s.	0.640**	n.s.	n.s.	n.s.	0.580*	n.s.	n.s.	-0.318+	n.s.	n.s.
0600 h Temp. °C,	(X_5)	-0.397+	-0.301+	-0.407+	-0.506+	-0.380+	-0.323+	-0.332+	-0.426++	n.s.	n.s.	0.283+	n.s.
1800 h Temp. °C,	(X_6)	n.s.	-.0440++	n.s.	-0.656**	n.s.	-0.410++	n.s.	-0.582*	-.0626**	n.s.	n.s.	n.s.
Sunshine h/d,	(X_7)	0.362+	n.s.	n.s.	n.s.	0.340+	0.308+	.354+	n.s.	n.s.	0.409++	n.s.	n.s.
Max Hum %,	(X_8)	-0.523*	0.424++	-0.587+	n.s.	-0530*	0.431++	-0.586*	n.s.	n.s.	n.s.	n.s.	n.s.
Min Hum %,	(X_9)	n.s.	n.s.	-0.585*	0.639**	n.s.	n.s.	-0.517*	0.652**	n.s.	n.s.	n.s.	0.420++

n.s. Means simple correlation coefficient is not significant at the 0.15 alpha level of significance.

** Significant at 1% probability level, * Significant at 5% probability level.

++ Significant at 10% probability level, + Significant at 15% probability level.

n   Number of data pairs used in calculation.

Wind speed did not show significant effect upon the studied production variables.

(Sawan et al., 1999)

Equations obtained from data of the fourth quarter explained more variations of flower, boll production and boll retention ratio. Evaporation, humidity and temperature are the principal climatic factors that govern cotton flower and boll production during the fourth quarter; since they were most strongly correlated with the dependent variables studied (Table 11) (Sawan et al. 1999).

Table 11. Significant simple correlation values between the climatic factors and flower, boll production, and boll retention ratio due to quarters periods of production stage for the combined data of the two seasons. (n =30)

Climatic factors		Flower				Boll				Ratio:Bolls/Flowers (100)			
		1st	2nd	3rd	4th	1st	2nd	3rd	4th	1st	2nd	3rd	4th
MaxTemp °C,	(X_1)	n.s.	n.s.	0.29^+	-0.48^{**}	n.s.	n.s.	0.38^{++}	-0.47^{**}	0.27^+	n.s.	n.s.	n.s.
Min Temp °C,	(X_2)	n.s.	n.s.	-0.35^{++}	n.s.	n.s.	n.s.	-0.28^+	n.s.	n.s.	n.s.	n.s.	n.s.
Max-Min °C,	(X_3)	-0.40^*	-0.30^+	0.59^{**}	-0.36^{++}	n.s.	-0.48^{**}	0.52^{**}	-0.38^{++}	-0.40^*	-0.47^{**}	n.s.	-0.28^+
Evapor. mm/d,	(X_4)	0.78^{**}	n.s.	0.32^{++}	-0.67^{**}	0.67^{**}	-0.51^{**}	n.s.	-0.74^{**}	n.s.	-0.82^{**}	-0.49^{**}	-0.72^{**}
0600 h Temp. °C,	(X_5)	n.s.	0.27^+	-0.43^*	-0.31^+	n.s.	n.s.	-0.37^{++}	-0.37^{++}	n.s.	n.s.	n.s.	n.s.
1800 h Temp. °C,	(X_6)	n.s.	n.s.	n.s.	-0.42^*	n.s.	n.s.	n.s.	-0.37^{++}	n.s.	n.s.	n.s.	n.s.
Sunshine h/d,	(X_7)	n.s.	n.s.	0.38^{++}	n.s.	n.s.	n.s.	0.32^{++}	n.s.	n.s.	0.30^+	n.s.	0.27^+
Max Hum %,	(X_8)	n.s.	n.s.	n.s.	-0.64^{**}	n.s.	n.s.	n.s.	-0.71^{**}	n.s.	-0.60^{**}	-0.44^*	-0.70^{**}
Min Hum %,	(X_9)	n.s.	n.s.	-0.54^{**}	0.69^{**}	-0.32^{++}	0.42^*	-0.37^{++}	0.72^{**}	n.s.	0.72^{**}	0.40^*	0.56^{**}
R^2		0.667	0.116	0.496	0.672	0.446	0.335	0.389	0.747	0.219	0.737	0.269	0.615

(Sawan et al., 1999)

Evaporation that seems to be the most important climatic factor (Sawan et al., 1999), and had a negative relationship which means that high evaporation ratio significantly reduces flower and boll production. Maximum temperature, temperature-differentiates and maximum humidity also showed negative links with fruiting production (Sawan et al. 1999), which indicates that these climatic variables have determinable effect upon Egyptian cotton fruiting production. Minimum humidity was positively high correlated in most quarter periods for flower, boll production and boll retention ratio (Sawan et al. 1999). This means that an increase of this factor will increase both flower and boll production. Maximum temperature is sometime positively and sometime negatively linked to boll production (Table 11) (Sawan et al. 1999). These erratic correlations may be due to the variations in the values of this factor between the quarters of the production stages, as shown from its range and mean values (Table 1) (Sawan et al. 2005).

Burke et al. (1990) pointed out that the usefulness of the 27.5°C midpoint temperature of the TKW of cotton as a baseline temperature for a thermal stress index (TSI) was investigated in field trials on cotton cv. Paymaster 104. This biochemical baseline and measurements of foliage temperature were used to compare the TSI response with the cotton field performance. Foliage temperature was measured with hand-held 4°C field of view IR thermometer while plant biomass was measured by destructive harvesting. The biochemical based TSI and the physically based crop water stress index were highly correlated ($r^2 = 0.92$) for cotton across a range of environmental conditions. Reddy et al. (1995) in controlled environmental chambers pima cotton cv. S-6 produced less total biomass at 35.5°C than at 26.9°C and no bolls were produced at the higher temperature 40°C. This confirms the results of this study as maximum temperature showed a negative relationship with production variables in the fourth quarter period of the production stage. Zhen (1995) found that the most important factors decreasing cotton yields in Huangchuan County, Henan, were low temperatures in spring, high temperatures and pressure during summer and the sudden fall in temperature at the beginning of autumn. Measures to increase yields included the use of the more suitable high-oil cotton cultivars, which mature early, and choosing sowing dates and spacing so that the best use was made of the light and temperature resources available.

It may appear that the grower would have no control over boll shedding induced by high temperature, but this is not necessarily the case. If he can irrigate, he can exert some control over temperature since transpiring plants have the ability to cool themselves by evaporation. The leaf and canopy temperatures of drought-stressed plants can exceed those of plants with adequate quantity of water by several degrees when air humidity is low (Ehrler 1973). The grower can partially overcome the adverse effects of high temperature on net photosynthesis by spacing plants to adequately expose the leaves. Irrigation may also increase photosynthesis by preventing stomata closure during the day. Adequate fertilization is necessary for maximum rates of photosynthesis. Finally, cultivars appear to differ in their heat tolerance (Fisher 1975). Therefore, the grower can minimize boll abscission where high temperatures occur by selecting a heat-tolerant cultivar, planting date management, applying an adequate fertilizer, planting or thinning for optimal plant spacing, and irrigating as needed to prevent drought stress.

Table 12. Significant simple correlation values between the climatic factors and flower, boll production and boll retention ratio for combined data of the two seasons (n = 120)

Climatic factors		Flower	Boll	Ratio
MaxTemp °C,	(X_1)	-0.152++	n.s.	n.s.
Min Temp °C,	(X_2)	n.s.	n.s.	n.s.
Max-Min °C,	(X_3)	-0.259**	-0.254**	n.s.
Evapor.mm/d,	(X_4)	-0.327**	-0.429**	-0.562**
0600 h Temp. °C,	(X_5)	n.s.	n.s.	n.s.
1800 h Temp. °C,	(X_6)	-0.204*	-0.190++	n.s.
Sunshine h/d,	(X_7)	-0.227*	-0.180++	n.s.
Max Hum %,	(X_8)	n.s.	n.s	-0.344**.
Min Hum %,	(X_9)	0.303**	0.364**	0.335**
R^2		0.406**	0.422**	0.336*

(Sawan et al.,1999)

3.4. Intervals of Days Required for Determining Efficient Relations between Climatic Factors and Cotton Flower and Boll Production

Multiple linear regression equations containing selected predictive variables were computed for the determined interval and coefficients of multiple determinations (R^2) were calculated to measure the success of the regression models in explaining the variation in data (Sawan et al., 2002b). Observation used in the statistical analysis were obtained during the entire production stage of flowering and boll development (60 days for each season, 29 June to 27 August).

Tables 13 and 14 (Sawan et al. 2002b) include the significant simple correlation coefficients between the production variables and the studied climatic factors for the different intervals of days in each season and the combined data for both seasons. All significant relationships were negative except that between maximum humidity in the first season, and minimum humidity in the two seasons which were positive.

Table 13. Significant simple correlation coefficient values between the production variables and the studied climatic factors for the different intervals of days in each season

Intervals of days	Production variables	Temp. °C Max. (X_1)	Temp. °C Min. (X_2)	Max.-Min. °C (X_3)	Evap. mm/day (X_4)	Surface soil temp. °C 0600 h (X_5)	Surface soil temp. °C 1800 h (X_6)	Sunshine duration h/day (X_7)	Humidity Max. (X_8)	Humidity Min. (X_9)
					First season					
2 Days ($n^{\#}$ = 30)	Flower	NS	NS	NS	-0.60^{**}	NS	NS	-0.30^{++}	0.33^{++}	0.21^{+}
	Boll	NS	NS	NS	-0.56^{**}	NS	NS	NS	0.33^{++}	NS
	Boll retention ratio	NS	NS	NS	-0.27^{++}	NS	NS	NS	NS	NS
3 Days ($n^{\#}$ = 20)	Flower	NS	NS	NS	-0.68^{**}	NS	NS	-0.32^{+}	0.39^{++}	0.29^{+}
	Boll	NS	NS	NS	-0.63^{**}	NS	NS	NS	0.36^{++}	NS
	Boll retention ratio	NS	NS	NS	-0.31^{+}	-0.24^{+}	NS	NS	NS	NS
4 Days ($n^{\#}$ = 15)	Flower	NS	NS	NS	-0.71^{**}	NS	NS	-0.35^{+}	0.36^{++}	0.25^{+}
	Boll	NS	NS	NS	-0.64^{**}	NS	NS	-0.23^{+}	0.38^{++}	NS
	Boll retention ratio	NS	NS	NS	-0.29^{+}	NS	NS	NS	NS	NS
5 Days ($n^{\#}$ = 12)	Flower	NS	NS	NS	-0.67^{*}	NS	NS	-0.41^{+}	0.45^{++}	NS
	Boll	NS	NS	NS	-0.61^{*}	NS	NS	NS	0.43^{+}	NS
	Boll retention ratio	NS	NS	NS	-0.30^{+}	-0.30^{+}	NS	NS	NS	NS
6 Days ($n^{\#}$ = 10)	Flower	NS	NS	NS	-0.73^{**}	NS	NS	-0.44^{+}	0.46^{+}	0.40^{+}
	Boll	NS	NS	NS	-0.69^{*}	NS	NS	NS	0.41^{+}	NS
	Boll retention ratio	NS	NS	NS	-0.36^{+}	-0.39^{+}	NS	NS	NS	NS
10 Days ($n^{\#}$ = 6)	Flower	NS	NS	NS	-0.79^{*}	NS	NS	-0.57^{++}	0.79^{*}	NS
	Boll	NS	NS	NS	-0.71^{++}	NS	NS	-0.40^{+}	0.71^{++}	NS
	Boll retention ratio	NS	NS	NS	-0.39^{+}	-0.35^{+}	NS	NS	NS	NS
					Second season					
2 Days ($n^{\#}$ = 30)	Flower	-0.47^{**}	NS	NS	-0.70^{**}	NS	-0.41^{*}	-0.44^{*}	NS	0.55^{**}
	Boll	-0.47^{**}	NS	NS	-0.70^{**}	NS	-0.41^{*}	-0.44^{*}	NS	0.55^{**}
	Boll retention ratio	NS	NS	-0.36^{*}	NS	NS	NS	NS	NS	0.22^{+}
3 Days ($n^{\#}$ = 20)	Flower	NS	NS	$-.45^{*}$	-0.71^{**}	NS	-0.47^{*}	-0.54^{*}	NS	0.55^{**}
	Boll	NS	NS	-0.46^{*}	-0.70^{**}	NS	-0.46^{*}	-0.55^{**}	NS	0.55^{**}
	Boll retention ratio	NS	NS	-0.28^{+}	-0.22^{+}	NS	NS	-0.32^{+}	NS	0.25^{+}
4 Days ($n^{\#}$ = 15)	Flower	NS	NS	-0.49^{++}	-0.68^{**}	NS	-0.49^{++}	-0.45^{++}	NS	0.54^{*}
	Boll	NS	NS	-0.50^{*}	-0.73^{**}	NS	-0.48^{++}	-0.53^{*}	NS	0.57^{*}
	Boll retention ratio	NS	NS	-0.41^{++}	-0.43^{++}	NS	NS	-0.54^{*}	NS	0.47^{++}
5 Days ($n^{\#}$ = 12)	Flower	NS	NS	-0.51^{++}	-0.81^{**}	NS	-0.58^{*}	-0.62^{*}	NS	0.65^{*}
	Boll	NS	NS	-0.51^{++}	-0.80^{**}	NS	-0.59^{*}	-0.61^{*}	NS	0.65^{*}
	Boll retention ratio	NS	NS	-0.40^{+}	NS	NS	-0.57^{*}	NS	NS	0.53^{++}
6 Days ($n^{\#}$ = 10)	Flower	NS	NS	-0.53^{++}	-0.78^{**}	NS	-0.62^{*}	-0.66^{*}	NS	0.61^{*}
	Boll	NS	NS	-0.55^{++}	-0.78^{**}	NS	-0.63^{*}	-0.66^{*}	NS	0.62^{*}
	Boll retention ratio	NS	NS	-0.80^{**}	-0.65^{*}	NS	NS	-0.66^{*}	NS	0.69^{*}
10 Days ($n^{\#}$ = 6)	Flower	NS	NS	-0.67^{++}	-0.39^{+}	NS	-0.52^{+}	-0.69^{++}	NS	0.67^{++}
	Boll	NS	NS	-0.64^{++}	-0.43^{+}	NS	-0.48^{+}	-0.70^{++}	NS	0.72^{++}
	Boll retention ratio	NS	NS	NS	-0.35^{+}	NS	-0.32^{+}	-0.34^{+}	NS	NS

** Significant at 1 % probability level, * Significant at 5 % probability level.
++ Significant at 10 % probability level, + Significant at 15 % probability level.
NS Means simple correlation coefficient is not significant at the 15% probability level.
n = Number of data pairs used in calculations.
Wind speed did not show significant effect upon the studied production variables.
(Sawan et al., 2002b)

Table 14. Significant simple correlation coefficient values between the production variables and the studied climatic factors for the different intervals of days combined over both seasons

Intervals of days	Production variables	Climatic factors								
		Temp. °C		Max.-Min. °C	Evap. mm/day	Surface soil temp. °C		Sunshine duration h/day	Humidity	
		Max. (X_1)	Min. (X_2)	(X_3)	(X_4)	0600 h (X_5)	1800 h (X_6)	(X_7)	Max. (X_8)	Min. (X_9)
2 Days ($n^{\#}$ = 60)	Flower	-0.31[++]	NS	-0.32[*]	-0.36[**]	NS	-0.24[+]	-0.36[**]	NS	0.37[**]
	Boll	-0.29[++]	NS	-0.30[++]	-0.46[**]	NS	-0.21[+]	-0.31[*]	NS	0.44[**]
	Boll retention ratio	NS	NS	NS	-0.61[**]	NS	NS	NS	-0.48[**]	0.40[**]
3 Days ($n^{\#}$ = 40)	Flower	-0.34[*]	NS	-0.34[*]	-0.33[*]	NS	-0.28[++]	-0.39[*]	NS	0.34[*]
	Boll	-0.32[*]	NS	-0.32[*]	-0.48[**]	NS	-0.24[+]	-0.36[*]	NS	0.45[**]
	Boll retention ratio	NS	NS	NS	-0.63[**]	NS	NS	NS	-0.53[**]	0.40[*]
4 Days ($n^{\#}$ = 30)	Flower	-0.31[++]	NS	-0.35[++]	-0.33[++]	NS	-0.28[+]	-0.39[*]	NS	0.34[++]
	Boll	-0.31[++]	NS	-0.33[++]	-0.48[**]	NS	-0.23[+]	-0.38[*]	NS	0.45[*]
	Boll retention ratio	NS	NS	NS	-0.64[**]	NS	NS	NS	-0.48[**]	0.42[*]
5 Days ($n^{\#}$ = 24)	Flower	-0.35[++]	NS	-0.37[++]	-0.39[++]	NS	-0.39[++]	-0.52[**]	NS	0.41[*]
	Boll	-0.33[+]	NS	-0.35[++]	-0.49[*]	NS	-0.35[++]	-0.44[*]	NS	0.47[**]
	Boll retention ratio	NS	NS	NS	-0.66[**]	NS	NS	NS	-0.56[**]	0.43[*]
6 Days ($n^{\#}$ = 20)	Flower	-0.37[++]	NS	-0.41[++]	-0.38[++]	NS	NS	-0.54[**]	NS	0.42[*]
	Boll	-0.37[++]	NS	-0.40[++]	-0.49[*]	NS	NS	-0.46[*]	NS	0.49[*]
	Boll retention ratio	NS	NS	NS	-0.69[**]	NS	NS	NS	-0.56[**]	0.45[*]
10 Days ($n^{\#}$ =12)	Flower	NS	NS	-0.45[++]	-0.40[+]	NS	-0.55[*]	-0.65[*]	NS	0.43[++]
	Boll	NS	NS	-0.43[++]	-0.51[++]	NS	-0.53[++]	-0.57[*]	NS	0.51[++]
	Boll retention ratio	NS	NS	NS	-0.74[**]	NS	NS	NS	-0.63[*]	0.55[*]

[**] Significant at 1 % probability level, [*] Significant at 5 % probability level.
[++] Significant at 10 % probability level, [+] Significant at 15 % probability level.
NS Means simple correlation coefficient is not significant at the 15% probability level.
[#] n = Number of data pairs used in calculation.
(Sawan et al., 2002b)

Evaporation was the most important climatic factor affecting flower and boll production in Egyptian cotton (Sawan et al. 2002b). The negative correlation means that high evaporation ratio significantly reduced flower and boll production. The second most important climatic factor was minimum humidity (Sawan et al. 2002b), which had a high positive correlation with flower and boll production, and retention ratio. The positive correlation means that increased humidity would bring about better boll production. The third most important climatic factor in this study was sunshine duration (Sawan et al. 2002b), which showed a significant negative relationship with flower and boll production only.

Maximum air temperature, temperature magnitude and surface soil temperature at 1800 h show significant negative relationships with flower and boll production only. Meanwhile, the least important factors were surface soil temperature at 0600 h and minimum air temperature (Sawan et al. 2002b).

Response of Flower and Boll Development to Climatic Factors ... 35

These results indicated that evaporation was the most effective climatic factor affecting cotton boll production (Sawan et al. 2002b). As the sign of the relationship was negative, this means that an increase in evaporation caused a significant reduction in boll number.

Comparing results for the different intervals of days with those from daily observation, the 5-day interval appeared to be the most suitable interval, which actually revealed a more solid and more obvious relationships between climatic factors and production characters (Sawan et al. 2002b). This was in fact indicated by the higher R^2 values obtained when using the 5-day intervals. The climatic factors and the production characters in 5-day interval may be the most suitable interval for diminishing the daily fluctuations between the factors under study to clear these relations comparing with the other intervals. However, it is worthwhile to mention that this concept is true provided that the fluctuations in climatic conditions are limited or minimal (Sawan et al. 2002b). Therefore, it would be the most efficient interval to use to help circumvent the unfavorable effect of climatic factors. This finding gives researchers and producers a chance to deal with condensed rather than daily weather data.

The main climatic factors from this study affecting the number of flowers and bolls, and by implication yield, are evaporation and humidity (water stress) with being by far the most important factor (Sawan et al. 2002b).

Comparing results of both seasons for the different intervals of days (Table 14), with those from daily observation (Table 15) (Sawan et al. 2002b), it evident that five day interval was the most efficient interval with adequate and sensible relation and highest R^2 values. It is also more convenient since it possessed less data pairs (n = 24). This finding gives researchers a chance to deal with condensed rather than daily measured data.

From the multiple regression analysis of the 5 days interval; separate equation for each of the cotton production variables was computed to clarify the effect of the climatic factors in the equation. The following are the multiple regression equation from 5 days interval for the studied cotton production variables (Sawan et al. 2002b),

1. Flowers production.
Y = 7352 + 14.87X1 - 18.17X3 - 12.86X4 - 44.22X6 - 32.61X7 - 3.639X9
With R^2 = 0. 537**

2. Boll production.
Y = 5886 + 16.74X1 - 18.32X3 - 14.02X4 - 41.79X6 - 19.43X7 - 3.329X9
With R^2 = 0. 541**

3. Boll retention ratio.

$Y = 188.1 - 0.975X4 - 0.096X8 - 0.190X9$

With $R^2 = 0.532**$

The R^2 values indicate the importance of such equations since the climatic factors (X's) explained about 55% of the variation in the dependent variables (Sawan et al. 2002b).

Table 15. Significant simple correlation coefficient values between production variables and the studied the climatic factors for combined daily data over both seasons (n# = 120)

Climatic factors		Flower	Boll	Boll retention ratio
Max. Temp. °C,	(X_1)	-0.152^{++}	NS	NS
Min. Temp. °C,	(X_2)	NS	NS	NS
Max.-Min. Temp. °C,	(X_3)	-0.259^{**}	-0.254^{**}	NS
Evap. mm/day,	(X_4)	-0.327^{**}	-0.429^{**}	-0.562^{**}
0600 h Temp. °C,	(X_5)	NS	NS	NS
1800 h Temp. °C,	(X_6)	-0.204^{*}	-0.190^{++}	NS
Sunshine h/d,	(X_7)	-0.227^{*}	-0.180^{++}	NS
Max. Hum.%,	(X_8)	NS	NS	-0.344^{**}
Min. Hum.%,	(X_9)	0.303^{**}	0.364^{**}	0.335^{**}

** Significant at 1 % probability level, * Significant at 5 % probability level.
++ Significant at 10 % probability level.
NS Means simple correlation coefficient is not significant at the 15% probability level.
n = number of data pairs used in calculations.
(Sawan et al., 2002b)

Wang and Whisler (1994) found that climatic factors resulting in maximum cotton yield at Mississippi were: maximum temperature -1%; minimum temperature was 0 to 5%; solar radiation was -10%; wind speed was -10 or +25% and rainfall was +1 inch. Gutiérrez and López (2003) studied the effects of heat on the yields of cotton in Andalucia, Spain, during 1991-98, and found that high temperatures were implicated in the reduction of unit production. There was a significant negative relationship between average production and number of days with temperatures greater than 40°C and the number of days with minimum temperatures greater than 20°C. Wise et al. (2004) indicated that restrictions to photosynthesis could limit plant growth at high temperature in a variety of ways. In addition to increasing photorespiration, moderately high temperatures (35-42°C) can cause direct injury to the photosynthetic apparatus. Both carbon metabolism and thylakoid reactions have been suggested as the primary site of injury at these temperatures.

CONCLUSIONS

Results obtained from this study indicate that evaporation, sunshine duration, humidity, surface soil temperature at 1800 h, and maximum temperature, were the most significant climatic factors affecting flower and boll production of Egyptian cotton (Sawan et al. 2002a). Also, it could be concluded that during the 15-day periods both prior to and after initiation of individual bolls, evaporation, minimum humidity and sunshine duration, were the most significant climatic factors affecting cotton flower and boll production and retention in Egyptian cotton (Sawan et al. 2005). The negative correlation between evaporation and sunshine duration with flower and boll formation along with the positive correlation between minimum humidity and flower and boll production, indicate that low evaporation rate, short periods of sunshine duration and high value of minimum humidity would enhance flower and boll formation (Sawan et al. 2005). Temperature appeared to be less important in the reproductive growth stage of cotton in Egypt than evaporation (water stress), sunshine duration and minimum humidity. These findings concur with those of other researchers except for the importance of temperature. A possible reason for that contradiction is that the effects of evaporation rate and humidity were not taken into consideration in the research studies conducted by other researchers in other countries (Sawan et al. 2005). Temperature and evaporation are closely related to each other to such an extent that the higher evaporation rate could possible mask the effect of temperature. Water stress is the main player and other authors have suggested means for overcoming its adverse effect which could be utilized for the Egyptian cotton. It must be kept in mind that although the reliable prediction of the effects of the aforementioned climatic factors could lead to higher yields of cotton, yet only 50% of the variation in yield could be statistically explained by these factors and hence consideration at the same time should be given to the management practices presently in use. In conclusion, the early prediction of possible adverse effects of climatic factors could pave the way for adopting adequate precautions regarding the effect of certain climatic factors on production of Egyptian cotton. This would be useful to minimize the deleterious effects of these factors, through the application of adequate management practices, i.e. adequate irrigation regime (Orgas et al. 1992), as well as utilization of specific plant growth regulators (Mosely et al. 1994; Zhao and Oosterhuis 1997; Meek et al. 1999) which would limit and control the negative effects of some climatic factors, and this will lead to an improvement in cotton yield in Egypt. Also, it could be concluded that the fourth quarter period of the production stage (Sawan et al. 1999) is the most appropriate and usable production time to collect data for determining efficient prediction equations for cotton flower and boll production in Egypt, and making valuable recommendations. The 5-day interval was found to give adequate and sensible relationships between climatic factors and cotton production growth under Egyptian conditions when compared with other intervals and daily observations (Sawan et al. 2002b). Collecting data at a 5-day interval will certainly reduce computational resources and give better results than obtained from the other intervals used in this study or daily records. Evaporation and sunshine duration appeared to be important climatic factors affecting boll production in Egyptian cotton. Our findings indicate that increasing evaporation rate and sunshine duration resulted in lower boll production. On the other hand, humidity, which had a positive correlation with boll production, was also an important climatic factor. In general, increased humidity would bring about better boll production. Finally, it may be concluded that the 5-day accumulation of climatic data during the production stage, in the absence of sharp fluctuations in these factors, could be used to forecast adverse effects on cotton production and with the application of appropriate production practices circumvent possible production shortage.

It could be stated that during the production stage, an accurate weather forecasting for the next 5-7 days would provide information to avoid any adverse effects of climatic factors on cotton production. It would be useful to minimize the deleterious effects of those factors through utilizing proper cultural practices which could limit and control their negative effects. This will lead to a useful improvement in cotton yielding.

REFERENCES

Barbour MM, Farquhar GD (2000). Relative humidity- and ABA-induced variation in carbon and oxygen isotope ratios of cotton leaves. *Plant Cell Environ* 23: 473-485.

Bhatt JG (1977) Growth and flowering of cotton (*Gossypium hirsutum* L.) as affected by daylength and temperature. *J of Agric Sci* 89: 583-588.

Boyer JS, Johnson RR, Saupe SG (1980). Afternoon water deficits and grain yields in old and new soybean cultivars. *Agron J* 72: 981-986.

Burke JJ, Hatfield JL, Wanjura DF (1990). A thermal stress index for cotton. *Agron J* 82: 526-530.

Burke JJ, Mahan JR, Hatfield JL (1988). Crop specific thermal kinetic windows in relation to wheat and cotton biomass production. *Agron J* 80: 553-556.

Draper NR, Smith H (1966). *Applied Regression Analysis*. Wiley, New York, 407 pp.

Ehrler WL (1973). Cotton leaf temperatures as related to soil water depletion and meteorological factors. *Agron J* 65: 404-409.

El-Zik KM (1980). *The cotton plant - its growth and development. Western Cotton Prod*. Conf. Summary Proc., Fresno, CA, pp 18-21.

Fisher WD (1975). Heat induced sterility in Upland cotton. In *Proceedings 27th cotton improvement conferences* 85.

Gipson LR, Joham HE. (1968). Influence of night temperature on growth and development of cotton (*Gossypium hirsutum* L.): I. Fruiting and boll development. *Agron J* 60: 292-295.

Guinn G (1982). *Causes of square and boll shedding in cotton*. USDA Tech. Bull. 1672. USDA, Washington, DC.

Guo Y, Landivar JA, Hanggeler JC, Moore J (1994). Response of cotton leaf water potential and transpiration to vapor pressure deficit and salinity under arid and humid climate conditions. In *Proceedings Beltwide cotton conferences*, San Diego, CA, USA, 5-8 January, National Cotton Council, Memphis, TN, USA, pp 1301-1308.

Gutiérrez Mas JC, López M (2003). Heat, limitation of yields of cotton in Andalucia. *Agricultura, Revista Agropecuaria* 72: 690-692.

Hearn AB, Constable GA (1984). *The physiology of tropical food crops*. In: Goldsworth PR; Fisher NM (eds) Chapter 14: cotton. Wiley, New York, pp 495-527 (664 pp).

Hodges HF, Reddy KR, McKinion JM, Reddy VR (1993). *Temperature effects on cotton*. Bulletin Mississippi Agricultural and Forestry Experiment Station, no 990, 15 pp.

Holaday AS, Haigler CH, Srinivas NG, Martin LK, Taylor JG (1997). Alterations of leaf photosynthesis and fiber cellulose synthesis by cool night temperatures. In *Proceedings Beltwide cotton conferences*, New

Orleans, LA, 6-10 January, National Cotton Council, Memphis, TN, USA, pp 1435-1436.

Human JJ, Du Toit D, Bezuidenhout HD, De Bruyn LP (1990). The influence of plant water stress on net photosynthesis and yield of sunflower (*Helianthus annuus* L.). *J Agron Crop Sci* 164: 231-241.

Kaur R, Singh OS (1992). Response of growth stages of cotton varieties to moisture stress. *Indian J Plant Physiol* 35: 182-185.

McKinion JM, Reddy KR, Wall GW, Bhattacharya NC, Hodges HF, Bhattacharya S (1991). Growth response of Pima cotton to CO_2 enrichment during vegetative period. In *Proceedings Beltwide cotton conferences*, San Antonio, TX, USA, 9-12 January, National Cotton Council, Memphis, TN, USA, 842 pp.

Meek CR, Oosterhuis DM, Steger AT (1999). Drought tolerance and foliar sprays of glycine betaine. In *Proceedings Beltwide cotton conferences*, Orlando, FL, USA, 3-7 January, National Cotton Council, Memphis, TN, USA, pp 559-561.

Mergeai G, Demol J (1991). Contribution to the study of the effect of various meteorological factors on production and quality of cotton (*Gossypium hirsutum* L.) fibers. *Bull Rech Agron Gembloux* 26: 113-124.

Miller JK, Krieg DR, Paterson RE (1996). Relationship between dryland cotton yields and weather parameters on the Southern Hig Plains. In *Proceedings Beltwide cotton conferences*, Nashville, TN, USA, 9-12 January, National Cotton Council, Memphis, TN, USA, pp 1165-1166.

Moseley, D., Landivar, J.A., Locke, D., (1994). Evaluation of the effect of methanol on cotton growth and yield under dry-land and irrigated conditions. In *Proceedings Beltwide cotton conferences*, San Diego, CA, USA, 5-8 January, National Cotton Council, Memphis, TN, USA, pp 1293-1294.

Oosterhuis DM (1997). Effect of temperature extremes on cotton yields in Arkansas. In: Oosterhuis DM, Stewart JM (eds) *Proceedings of the cotton research meeting*, held at Monticello, Arkansas, USA, 13 February. Special Report-Arkansas Agricultural Experiment Station, Division of Agriculture, University of Arkansas, no. 183, pp 94-98.

Oosterhuis DM (1999). Yield response to environmental extremes in cotton. In: *Proceedings of the cotton research meeting*, Fayetteville, USA. Special Report - Arkansas Agricultural Experiment Station, no 193. Arkansas Agricultural Experiment Station, University of Arkansas, pp 30-38.

Orgaz F, Mateos L, Fereres E (1992). Season length and cultivar determine the optimum evapotranspiration deficit in cotton. *Agron J* 84: 700-706.

Reddy KR, Davidonis GH, Johnson AS, Vinyard BT (1999).Temperature regime and carbon dioxide enrichment alter cotton boll development and fiber properties. *Agron J* 91: 851-858.

Reddy KR, Doma PR, Mearns LO, Boone MYL, Hodges HF, Richardson AG, Kakani VG (2002). Simulating the impact of climate change on cotton production in the Mississippi Delta. *Clim Res* 22: 271-281.

Reddy KR, Hodges HF, McKinion JM (1993). Temperature effects on Pima cotton leaf growth. *Agron J* 85: 681-686.

Reddy KR, Hodges HF, McKinion JM (1995). Carbon dioxide and temperature effects on pima cotton growth. *Agric Ecosyst Environ* 54: 17-29.

Reddy KR, Hodges, H.F., McKinion, J.M., (1996). Can cotton crops be sustained in future climates? In *Proceedings Beltwide cotton conferences*, Nashville, TN, USA, 9-12 January, National Cotton Council, Memphis, TN, USA, pp 1189-1196.

SAS Institute, Inc. (1985). *SAS user's guide: statistics*. 5th edn. SAS Institute, Cary, pp 433-506.

Sawan ZM, Hanna LI, Gad El Karim GhA, McCuistions WL (2002a). Relationships between climatic factors and flower and boll production in Egyptian cotton (*Gossypium barbadense*). *J Arid Environ* 52: 499-516.

Sawan ZM, Hanna LI, McCuistions WL (1999). Effect of climatic factors during the development periods of flowering and boll formation on the production in Egyptian cotton (*Gossypium barbadense*). *Agronomie* 19: 435-443.

Sawan ZM, Hanna LI, McCuistions WL (2002b). Intervals of days required for determining efficient rlations between climatic factors and cotton flower and boll production. *Can J Plant Sci* 82: 499-506.

Sawan ZM, Hanna LI, McCuistions WL (2005). Response of flower and boll development to climatic factors before and after anthesis in Egyptian cotton. *Clim Res* 29: 167-179.

Schrader SM, Wise RR, Wacholtz WF, Ort DR, Sharkey TD (2004). Thylakoid membrane responses to moderately high leaf temperature in Pima Cotton. *Plant Cell Environ* 27: 725-735.

Wang X, Whisler FD (1994). Analyses of the effects of weather factors on predicted cotton growth and yield. *Bull Mississippi Agric. For Exp Station*, no 1014, p 51.

Ward DA, Bunce JA (1986). Responses of net photosynthesis and conductance to independent changes in the humidity environments of the upper and lower surfaces of leaves of sunflower and soybean. *J Exp Bot* 37: 1842-1853.

Wise RR, Olson AJ, Schrader SM, Sharkey TD (2004). Electron transport is the functional limitation of photosynthesis in field-grown Pima cotton plants at high temperature. *Plant Cell Environ* 27: 717-724.

Yuan J, Shi YJ, Pan ZX, Liu XL, Li CQ (2002). Effect of meteorological conditions on cotton yield in arid farmland. *China Cotton* 29: 10-11.

Zhao DL, Oosterhuis D (1997). Physiological response of growth chamber-grown cotton plants to the plant growth regulator PGR-IV under water-deficit stress. *Environ Exp Bot* 38: 7-14.

Zhao YZ (1981). Climate in Liaoning and cotton production. *Liaoning Agricultural Science*, no. 5, pp 1-5.

Zhen JZ (1995). *The causes of low cotton production in Huanghchuan County and measures to increase production*. Henan Nongye Kexue, no 2, pp 5-6.

Zhou ZG, Meng YL, Shi P, Shen YQ, Jia ZK (2000). Study of the relationship between boll weight in wheat-cotton double cropping and meteorological factors at boll-forming stage. *Acta Gossypii Sinica* 12: 122-126.

In: Flowering Plants
Editor: Jeremy J. Tellstone

ISBN: 978-1-61324-653-5
© 2011 Nova Science Publishers, Inc.

Chapter 2

RISK ASSESSMENT OF INORGANIC AND ORGANIC POLLUTANTS IN FLOWERING PLANTS

Simona Dobrinas, Alina Daria Soceanu and Gabriela Stanciu

Department of Chemistry, University "Ovidius" of Constantza,
124 Mamaia Blvd., 900527, Constantza, Romania

ABSTRACT

The first step in the risk assessment process is to identify potential health effects that may occur from different types of pollutants exposure. Humans are exposed to pollutants by different routes of exposure such as inhalation, ingestion and dermal contact. Heavy metals are very harmful because of their non-biodegradable nature, long biological half-lives and their potential to be accumulated in different body parts. Organochlorine pesticides (OCPs) were used for the first time in Romania in 1948. Since 1988 these kinds of products were banned or restricted in Romania and in the present only chlorinated insecticides on the base of lindane are used for seeds treatment in Romania, but this substance is not included in the Stockholm Convention list on Persistent Organic Pollutants. Human exposure to polycyclic aromatic hydrocarbons (PAHs) can occur through different environmental pathways, including internal absorption through food and water consumption. Human exposure to PAHs is 88-98% connected with food (in 5% with food of plant origin). In this study levels of heavy metals (Pb, Cd, Cu, Zn, Mn, Fe), OCPs (lindane, p,p'- DDT, p,p'- DDE, p,p'- DDD, HCB, aldrin, dieldrin, endrin and hepthaclor) and PAHs (Np, Acy, Ace, F, Ph, An, Fl, Py, B[a]An, Chry, B[k]Fl, B[a]Py, B[ghi]P, dB[a,h]An, I[1,2,3-cd]Py) in flowering plants from Solanaceae and Rosaceae families were investigated. Studied plants at various growing stages were: tomato and bell pepper plants at 5, 15 and 35 cm (samples were taken from roots, stems, leaves and fruits) and peach, nectarine, apricot, cherry, sour cherry, apple and quince trees (green, almost ripe and ripe fruits) from Romania's urban and rural areas. For metals determination was used a AA6200 Schimadzu FAAS. Analysis of PAHs was carried with a HP 5890/5972 GC-MS system and analysis of OCPs was carried with a HP 5890 gas chromatograph equipped with an electron capture detector. Repeatability of methods, expressed as the relative standard deviation, was lower than 7.5% while recoveries were in the range of 96–99%. The estimated provisional tolerable weekly intake (PTWI) of all studied metals was calculated. All pesticides and PAHs concentrations of analyzed flowering plants were compared with values imposed by European Communities regulations. According to ANOVA test, statistically significant differences were found between samples from urban and rural areas, respectively, among samples within each botanical origin. Data analysis of health risk estimates indicated that analyzed persistent organic pollutants do not pose a direct hazard to human health.

1. INTRODUCTION

The occurrence in the environment of toxic and persistent substances posing a risk for human health and ecosystems is a relevant environmental problem. Data from analytical and environmental chemistry play often a critical role since they are the raw information used to assess the distribution of pollutants in the environment and the associated potential risk for human health and ecosystems. It is therefore extremely difficult to produce evidence at the epidemiological level of a direct link between exposure to one and/or other of the chemical substances or products and the development of specific diseases. These substances or chemicals are more and more numerous: polycyclic aromatic hydrocarbons (PAHs), organo-halogenated derivatives, toxic metals, organochlorine pesticides (OCPs), food additives and others. Some of them are persistent in the environment and contaminate the air, water, soil and food, so people are constantly exposed to persistent toxic substances or products, including Persistent Organic Pollutants (POPs) [1, 2].

1.1. Heavy Metals

Food contamination with heavy metals is one of the most important aspects of food quality assurance. Heavy metals are non-biodegradable and persistent environmental contaminants, which may be deposited on the surfaces and then absorbed into tissues of vegetables. Plants take up heavy metals by absorbing them from deposits on parts of the plants exposed to the air from polluted environments, contaminated soils as well as due to irrigation with contaminated water. The pollution of the environment with toxic metals is a result of many human activities, such as mining and metallurgy, and the effects of these metals on the ecosystem are of large economic and public health significance.

Prolonged consumption of unsafe concentrations of heavy metals through foodstuffs may lead to the chronic accumulation of heavy metals in the kidney and liver of humans causing disruption of numerous biochemical processes and leading to cardiovascular, nervous, kidney and bone diseases [3, 4].

Lead is a naturally occurring element that is hazardous when present at elevated concentration. The most important sources of lead exposure are industrial emissions, soils, car exhaust gases and contaminated foods. The entrance of lead at levels >0.5–0.8 lg/mL into blood causes various abnormalities. Lead accumulates in the skeleton, especially in bone marrow. It is a neurotoxin and causes behavioral abnormalities, retarding intelligence and mental development. It interferes in the metabolism of calcium and vitamin D and affects hemoglobin formation and causes anemia.

Cadmium is a heavy metal, which is classified as a human carcinogen and is known to be toxic to plants. Cadmium ions are easily absorbed by vegetables and, in animal-based food, are principally distributed in the liver and kidneys.

Manganese is a plant micronutrient that, depending on its content in the soil and on factors that control its availability such as pH, organic matter and microbial activity, can achieve levels which are toxic for the plants. The critical level of foliar Mn concentration in plants to produce toxicity symptoms varies among species and even cultivars. Manganese is an essential nutrient that plays an important role in human health. Everyone comes in contact with small amounts of manganese in air, water and food. Long-term exposure to manganese at very high levels may result in permanent neurological (brain and central nervous system) damage.

Copper is a component of several important enzymes in the body tissues and is essential to good health. Its toxicity is a much overlooked contributor to many health problems: including anorexia, fatigue, premenstrual syndrome, depression, anxiety, migraine headaches, allergies, childhood hyperactivity and learning disorders.

Knowledge about Zn toxicity in humans is scarce. The most important information reported is its interference with Cu metabolism. The symptoms that an acute oral Zn dose may provoke include: tachycardia, vascular shock, dyspeptic nausea, vomiting, diarrhea, pancreatitis and damage of hepatic parenchyma.

Iron is critical for cholorphyll formation, photosynthesis and is also used by enzymes to regulate transpiration in plants. Too much iron can be bad, however. It competes with zinc and copper in the soil for position (availability) to plants. Iron is only required in very small amounts which is lucky because iron is usually only available in very small amounts in the soil. Normally, iron deficiency (lack of iron) is typically not a problem because the small minute amounts needed are typically readily available in soil. However, in some cases, when plants are grown on soil with high levels of calcium, iron can be tied in the soil and then iron deficiency symptoms may show in plants. Typically this manifests itself through the bottom leaves turning yellow, stunted growth, and the plants not responding to regular fertilizer, etc. When iron is deficient in tomatoes new foliage is small and yellowish, although green along veins and dead spots may develop between veins or on leaf tips. For humans, iron is a mineral present in certain enzymes and hemoglobin, the substance in red blood cells that enables the blood to transport oxygen throughout the body. The RDA (Recommended Dietary Allowance) for iron for the adult male is 10 mg/day, while that for the adult woman is 15mg/day [5-8].

1.2. Organochlorine Pesticides

Humans are exposed to pesticides (found in soil, water, air and food by different routes of exposure such as inhalation, ingestion and dermal contact. Exposure to pesticides results in acute and chronic health problems vary with the extent of exposure. Increasing incidence of cancer, chronic kidney diseases, suppression of the immune system, sterility among males and females, endocrine disorders, neurological and behavioral disorders, especially among children, have been attributed to chronic pesticide poisoning. Moderate human health hazards from the misapplication of pesticides include mild headaches, flu, skin rashes, blurred vision and other neurological disorders while rare, but severe human health hazards include paralysis, blindness and even death. Pesticide pollution to the local environment also affects the lives of birds, wildlife, domestic animals, fish and livestock.

OCPs such as DDT, hexachlorocyclohexane, aldrin and dieldrin, are among the most commonly used pesticides because of their low cost and versatility against various pests. Because of their potential for bioaccumulation and biological effects, these compounds were banned in developed nations a long time ago.

Due to their persistence, all these compounds can be still found in the global environment, where they cycle between soil, vegetation and air. Many of them are semivolatile, which makes them susceptible to long-range transport and wide geographical distribution. Due to their hydrophobicity and lipophilicity, they accumulate in soils, sediments, and in the fatty tissues of living organisms, they are a subject of biomagnifications in food chains and eventually toxic to humans and wildlife [9].

According to the Food and Agriculture Organization inventory [10], more than 500,000 tons of unused and obsolete pesticides are threatening the environment and public health in many countries.

The Stockholm Convention on POPs was adopted in May 2001 with the objective of protecting human health and the environment from the potential risks of persistent organic pollutants. Romania has ratified Stockholm Convention by the Law no 261 of June 2004 and becomes a Party of the Convention on 28 October 2004.

As a part of the convention, Romania elaborated National Implementation Plans, document which has been send to the Stockholm Convention Secretariat on April 2006. The overall objective of the National Implementation Plans is to reduce or eliminate releases from the existing stockpiles and wastes; to eliminate production of POPs; to restrict the use of DDT and to reduce unintentionally releases of Dioxins, HCB and PCBs from the social and economic activities.

For the first time in Romania OCPs were used in 1948. The first products were based in principal on DDT but it had been also used other products based on endrin, dieldrin, aldrin, heptachlor, chlordane and toxafen. Since 1988 these kinds of products are banned or restricted in Romania and in the present only chlorinated insecticides on the base of lindane are used for seeds treatment; lindane is not included in the Stockholm Convention list [11, 12].

1.3. Polycyclic Aromatic Hydrocarbons

PAHs are environmental pollutants, with a chemical structure of two or more fused aromatic rings. The US EPA identified a list of 16 PAHs as priority pollutants to be controlled due to the effects these might have on the human health and the environment Since all PAHs do no fulfill accomplish with the requirements of persistence, bioaccumulation, toxicity, and long-range transport, they are not considered POPs. In consequence, PAHs were not included on the list in the Stockholm Convention on POPs.

PAHs are formed as byproducts of incomplete combustion processes. Primary natural sources of airborne PAHs are forest fires and volcanoes. Chlorinated compounds, residential burning of wood, the use of coal in domestic burning, power generation, incineration and petrochemical manufacturing have been pointed out as the most significant pathways of PAH entrance into the environment.

PAH volatility is highly dependent on the molecular weight. In the atmosphere, light PAHs are frequently found free due to their low vapor pressure, whereas larger molecular weight compounds are mostly adsorbed onto particular matter. In aquatic compartments, in spite of their low solubility, only low molecular weight PAHs can sometimes be found dissolved in water. In sediments, biodegradation is the most important PAH degradation pathway. In spite of all the above, PAHs with 4 or more rings present a long persistence in the environment in general, and in soils and sediments in particular. Not only lower molecular weight, but also heavy particle-bound PAHs, can travel long-distances and deposit far away from where they are released.

Health effects of PAHs have been largely studied, especially in the last years. However, the complete knowledge about PAH toxicity is still uncertain, since every single compound may show different human health effects. Benzo[a]pyrene (B[a]P) has been, until now, the individual compound which more attention has received. Numerous studies have concluded that exposure to B[a]P and PAH mixtures containing this compound could induce cancerigen effects. Epidemiological studies have detected an increase in lung cancer in humans occupationally exposed to coke oven emissions and cigarette smoke, presumably containing large amounts of PAHs [13].

The present work provides data regarding levels of heavy metals (Pb, Cd, Cu, Zn, Mn, Fe), organochlorine pesticides (lindane, p,p'- DDT, p,p'- DDE, p,p'- DDD, HCB, aldrin, dieldrin, endrin and hepthaclor) and PAHs acenaphthene (Ace), acenaphthylene (Acy), fluorene (F), naphthalene (Np), anthracene (An), fluoranthene (Fl), phenanthrene (Ph), benzo[α]anthracene

(B[α]An), benzo[k]fluoranthene (B[k]Fl), chrysene (Chry), pyrene (Py), benzo[ghi]perylene (B[ghi]Pe), benzo[α]pyrene (B[α]Py), dibenzo[α,h]anthracene dB[α,h]An, indeno[1,2,3–cd]pyrene (I[1,2,3–cd]Py) in flowering plants from Solanaceae and Rosaceae families. Studied plants at various growing stages were: tomato and bell pepper plants at 5, 15 and 35 cm (samples were taken from roots, stems, leaves and fruits) and peach, nectarine, apricot, cherry, sour cherry, apple and quince trees (green, almost ripe and ripe fruits) from Romania's urban and rural areas.

1.5. Flowering Plants

Solanaceae, family of flowering plants (order Solanales) has 102 genera and nearly 2,500 species, many of considerable economic importance as food and drug plants. Among the most important of these are the potato (*Solanum tuberosum*); eggplant (*S. melongena*); tomato (*Lycopersicon esculentum*); garden, or capsicum, pepper (*Capsicum annuum* and *C. frutescens*); tobacco (*Nicotiana tabacum*); deadly nightshade, the source of belladonna (*Atropa belladonna*); the poisonous jimsonweed (*Datura stramonium*) and nightshades (*S. nigrum, S. dulcamara*, and others); and many garden ornamentals, such as the genera *Petunia, Lycium, Solanum, Nicotiana, Datura, Salpiglossis, Browallia, Brunfelsia, Cestrum, Schizanthus, Solandra, Streptosolen,* and *Nierembergia.*

Members of the Solanaceae family are found throughout the world but are most abundant and widely distributed in the tropical regions of Latin America, where about 40 genera are endemic. Very few members are found in temperate regions, and only about 50 species are found in the United States and Canada combined. The genus *Solanum* contains almost half of all the species in the family, including all the species of wild potatoes found in the Western Hemisphere. The poisonous alkaloids present in some species of the family have given the latter its sombre vernacular name of "nightshade."

Solanum lycopersicum is a plant of the nightshade family (Solanaceae). The tomato was introduced to Europe by the Spanish in the early 16th century. The Italians called the tomato *pomodoro* ("golden apple"), which has given rise to speculation that the first tomatoes known to Europeans were yellow; similarly, it has been suggested that the French called it *pomme* d'amour ("love apple") because it was thought to have aphrodisiacal properties. In France and northern Europe the tomato was at first grown as an ornamental plant. Since botanists recognized it as a relative of the poisons belladonna and deadly nightshade, it was regarded with suspicion as a food. (The roots and leaves of the tomato plant are in fact poisonous; they contain the neurotoxin solanine).

The genus *Capsicum* is a member of the Solanaceae family that includes tomato, potato, tobacco and petunia. The genus *Capsicum* consists of approximately 22 wild species and five domesticated species: *C. annuum*, *C. baccatum*, *C. chinense*, *C. frutescens*, and *C. pubescens*. (Bosland, 1994) *Capsicum annuum* L. is a herbaceous annual that reaches a height of one meter and has glabrous or pubescent lanceolate leaves, white flowers, and fruit that vary in length, color, and pungency depending upon the cultivar. Native to America, this plant is cultivated almost exclusively in Europe and the United States [14].

The rose family (*Rosaceae*), in the order Rosales, is a large plant family containing more than 100 genera and 2,000 species of trees, shrubs, and herbs. This family is represented on all continents except Antarctica, but the majority of species are found in Europe, Asia, and North America.

Most species in the Rosaceae have leaves with serrated margins and a pair of stipules where the leaf joins the stem. The majority of tree-sized arborescent species have leaves that are simple except for species of mountain ash (*Sorbus* spp.), which have compound leaves divided into five to seven leaflets. Conversely, most woody shrubs and herbs have compound leaves which are composed of three to 11 leaflets. Branch spines and prickles are common on trees and shrubs in the rose family. However, there is variability in the appearance of these structures even among species which occur in very similar habitats. For example, blackbrush, (*Coleogyne ramosissima*), a species found in pinion-juniper woodlands in the American Southwest, has long spines on which it bears flowers, while Apache plume (*Fallugia paradoxa*, is found in the same region and habitat but has no spines. On a much larger scale, trees of the genus *Crataegus*, which are collectively called thornapples or hawthorns, have prominent branch spines while most species of *Malus* and *Prunus* are without spines. Herbaceous species typically lack spines or prickles. Flowers in this family are typically radially symmetrical flat discs (actinomorphic) and contain both male and female floral structures in a single flower. Flower ovaries may be positioned below the sepals and petals (inferior) or above them (superior). In flowers having an inferior ovary, the carpels are surrounded by a hollow receptacle. Flowers typically have five sepals, five petals, numerous stamens and one to 50 carpels. Carpels in this family tend to remain free instead of becoming fused into a many chambered, single carpel. Anthers have two chambers, called locules, which split lengthwise to release thousands of pollen grains. Another distinguishing feature of flowers in this family is the presence of a structure called the epicalyx. The epicalyx is composed of five sepal-like structures which occur below and alternate with the true calyx.

Most species have large white, pink, or red petals which are designed to attract pollinating insects. Many white and pale pink flowers also produce volatile esters, chemicals which we perceive as pleasant odors, but are produced to attract insects. The chief pollinators of rose flowers are bees ranging in size from tiny, metallic green flower bees of the genus *Augochlora*, through honey bees (*Apis*), to large bumble bees (*Bombus*). These pollinators are unspecialized and also pollinate many other species which have actinomorphic flowers and offer copious pollen as a reward for flower visitation [15, 16].

Prunus avium (cherry), *Prunus cersus* (sour cherry) and *Prunus armeniaca* (apricot) belong to the same family: *Rosaceae*, *Prunoidae* subfamily and *Prunus* genera. *Prunoidae*, also called *Amygdaloideae* is the flowering plant subfamily. The fruit of these plants are known as stone fruit (botanically, a drupe), as each fruit contains a single, hard-shelled seed called a stone or pit [17].

The apple is a tree and its pomaceous fruit, of the species Malus domestica in the rose family Rosaceae. It is one of the most widely cultivated tree fruits. It is a small deciduous tree reaching 5-12 m tall, with a broad, often densely twiggy crown. The leaves are alternately arranged, simple oval with an acute tip and serrated margin, slightly downy below, 5-12 cm long and 3-6 cm broad on a 2-5 cm petiole. The flowers are produced in spring with the leaves, white, usually tinged pink at first, 2.5-3.5 cm diameter, with five petals. The fruit matures in autumn, and is typically 5-9 cm diameter (rarely up to 15 cm).

The quince Cydonia oblonga is the sole member of the genus Cydonia and native to warm-temperate southwest Asia in the Caucasus region. It is a small deciduous tree, growing 5-8 m tall and 4-6 m wide, related to apples and pears, and like them has a pome fruit, which is bright golden yellow when mature, pear-shaped, 7-12 cm long and 6-9 cm broad. The immature fruit is green, with dense grey-white pubescence which mostly rubs off before maturity in late autumn when the fruit changes color to yellow with hard flesh that is strongly perfumed. The leaves are alternately arranged, simple, 6-11 cm long, with an entire margin and densely pubescent with fine white hairs. The flowers, produced in spring after the leaves, are white or pink, 5 cm across, with five petals [18].

2. Materials and Methods

2.1. Reagents and Solutions

All metal stock solutions (1000 mg/L) were prepared by dissolving the appropriate amounts of the spectral pure metals in dilute acids (HNO_3) 1:1 and then diluting them with deionized water. The working solutions were prepared by diluting the stock solutions to appropriate volumes. The nitric acid 65% and hydrogen peroxide 25% solutions used were of ultra pure grade, purchased from Merck. All reagents were of analytical-reagent grade and all solutions were prepared using deionized water.

Standards of pesticides: HCB, Lindane, p,p'- DDT, p,p'- DDE, p,p'-DDD, Aldrin, Dieldrin, Endrin, Hepthaclor and standards of PAHs: Np, Acy, Ace, F, Ph, An, Fl, Py, B[a]An, Chry, B[k]Fl, B[a]Py, B[ghi]P, dB[a,h]An, I[1,2,3-cd]Py were supplied by International Atomic Energy Agency, Monaco laboratory.

For clean up step were used two usual sorbent materials of variable polarities (activated at 420°C for 4h before use): silica (0.2 – 0.5 mm) and aluminium oxide 90 (0.063 – 0.200 mm) supplied by Merck, Darmstadt, Germany for PAHs determination. Florisil was assayed for preconcentration step as sorbent material of variable polarities for pesticides determination. It (60 – 100 mesh) was obtained from Fluka (packed in Switzerland) and was activated overnight (12h) at 130°C before use. Anhydrous sodium sulphate (granulated for residue analysis) was activated at 200°C for 2h before use. Hexane, supplied by Merck, Darmstadt, Germany and dichlormethane supplied by J.T. Baker, were used as eluents.

2.2. Sampling

Samples of tomatoes and peppers were collected from rural (*Slava Rusa*) and urban (*Constanta*) gardens from Romania (figure 1). The roots, stems, leaves and fruits of studied plants were collected in different stages of plant development (5, 15 and 35 cm of plant growing). Slava Rusa is a village from Tulcea district (about 22 Km far away from the first city), situated near the Babadag forest and Constanta is the largest Romanian seaport on the Black Sea. Slava Rusa has a population of approximately 1350 and Constanta 310000.

Also, samples of fruits in different stages of growing (green, almost ripe and ripe fruits) like cherries, sour cherries and apricots from urban (*Cernavoda*) and rural (*Mereni*) areas, apples and quinces from *Crisan* (rural area), peaches and nectarines from Mereni (rural areas) were investigated.

The town Cernavoda is a Danube fluvial port, and it houses the Cernavodă Nuclear Power Plant, consisting of two CANDU reactors providing about 18% of Romania's electrical energy output. Mereni is a commune in Constanța County, about 40 Km far from Constanta town, while Crisan is a fishing village situated in Tulcea district on Sulina canal in Danube Delta. The samples were brought to the laboratory and processed further for analysis. First were properly washed to remove surface dust and were kept frozen at -15°C until analysis.

Figure 1. Urban and rural areas studied from Romania.

2.3. Sample Mineralisation

To determine the metals concentration, a mineralization step is recommended even for liquid or water-soluble foodstuffs, the destruction of the organic matter preventing both spectral interferences and the accumulation of the residues in the burner head and spray chamber. In this context analyzed samples were submitted digestion with 8 mL HNO_3 and 10 mL H_2O_2 at 150°C in a Digesdhal device provided by Hach Company. After the complete digestion samples solution was filtered, made up to 50 mL with deionized water and Pb, Cd, Cu, Zn, Mn and Fe were determined by FAAS in air/acetylene flame using an aqueous standard calibration curve. Analyses were made in triplicate and the mean values are reported.

2.4. Sample Extraction and Clean-up

For OCPs and PAHs analysis approximately, 10 g of each sample type was used for Soxhlet extraction over a period of 12 h with hexane as solvent. The solvent was removed under reduced pressure at 40° C. All OCPs and PAHs were quantified from the Soxhlet extract obtained in triplicate. The extracts were evaporated under vacuum using a rotary evaporator and then the concentrated extract was purified by column chromatography.

A home-made glass column containing a piece of glass wool on a glass frit was filled with 5 g of activated aluminium oxide, 5 g of activated silica-gel and about 1 g of anhydrous sodium sulfate on the top to fractionate the aliphatic and aromatic fractions. The sorbent was pre-washed with n-hexane as described in EPA methods 8270C and 3600C [19]. For OCPs fractions was used a home-made glass column containing a piece of glass wool on a glass frit filled with 5 g of Florisil and about 1 g of anhydrous sodium sulfate on the top. The PAHs and OCPs residues were eluted with n hexane: dichlormethane (3:1) mixture and the eluate was collected in a conical evaporating flask. The sorbent was not allowed to dry during the conditioning and sample loading steps. The eluate was finally concentrated in a Kuderna–Danish concentrator to approximately 1 mL and the concentrated aliquots were blown down under a gentle stream of nitrogen gas. The final volumes were injected.

2.5. Instrumental Analysis

2.5.1. F-AAS

A flame atomic absorption spectrometer Shimadzu AA6500 was used for the determination of heavy metals (Pb, Cd, Cu, Zn, Mn and Fe). An air-acetylene flame was used for all elements. Monoelement hollow cathode lamps were employed to measure the elements. The acetylene was of 99.999%

purity at a flow rate 1.8-2.0 L/min. The characteristics of metal calibration are presented in Table 1.

2.5.2. F-AAS Method Analytical Performance

The calibration curves were obtained using solutions prepared by diluting stock solutions with deionized water to the desired concentrations.

The detection and quantification limits were evaluated from calibration curves. LOD and LOQ values for each calibration line were obtained and calculated using the equations: $(3 \cdot s_a \text{-} a)/b$ and $(10 \cdot s_a \text{-} a)/b$, respectively, where b is the slope of the calibration curve and s_a is the standard deviation of intercept of regression equation [20]. Good linearity was observed with coefficients of determination, R^2, exceeding 0.9991. The LOQ values varied from 0.009 to 0.63 mg/Kg.

Table 1. Characteristics of metal calibration curves

Metal	λ, nm	Concentration range (ppm)	Correlation coefficient
Fe	248.3	0.020-4.000	0.9998
Mn	279.5	0.08-1.28	0.9999
Cd	228.8	0.04–0.64	0.9999
Zn	213.9	0.032-0.512	0.9999
Cu	324.7	0.1–1.200	0.9991
Pb	283.3	0.02-2.4	0.9999

The working concentration range was established by analyzing the lowest and the highest concentration values of the proposed concentration range, ten times each of them. The test of homogeneity variance was applied for these values.

As a test of variance in homogeneity, the F test was applied in order to evaluate significance differences of concentration range limits [21]. F test was undertaken to evaluate the regression and lack of fit significances [22]. The calibration is considered suitable if F is less than the one-tailed tabulated value (F_{tab}) at a P selected confidence level. The P value was determined (PG = s_1^2/s_2^2 for $s_1^2 > s_2^2$ and PG = s_2^2/s_1^2 for $s_2^2 > s_1^2$) [21] and compared with F value for n-1=9 free degrees to evaluate significance differences. The calculated P values were compared with F_{tab} value (5.35 for n-1=9 free degrees) and obtained results showed that the calculated P values was below the F_{tab} value meaning that for our working ranges (see table 1) no significant differences were found between the variances of the concentration range limits. So, that the working ranges were correctly choosen.

Precision of the F-AAS method was determined by studying the repeatability (indicated precision under the same operating conditions over a short interval of time). The precision was successfully demonstrated by achieving RSD% values less than 7.5%, indicating that RSD are below those given by the Horwitz equation. These results showed that the current method is precise.

Accuracy of the method was evaluated using the standard addition method and was determined on samples at spiking levels of 0.01-0.05 mg/kg from the metals standard solutions. The recovery studies were carried out in triplicate for replicates of spiked samples. Results were satisfactory, with recoveries between 96 and 102% indicating the high accuracy of the method. These percentage recoveries ranged within the limits imposed by the Horwitz equation (85-110%) for the established concentration range [23]. These results reveal that Pb, Cd, Cu, Zn, Mn and Fe could be accurately determined by the F AAS method.

2.5.3. GC-ECD and GC-MS

A Hewlett-Packard 5890 gas chromatograph (GC) equipped with an electron capture detector (ECD) and a HP–5 fused–silica capillary column (30m×0.32mm×0.25µm) has been used for OCP analysis. Operating conditions were as follows: initial temperature 60°C (1 min), increased at a rate of 20°C/min to 300°C and finally held for 10 min; injector temperature: 250°C; detector temperature: 300°C; carrier gas: He; column flow-rate: 1.36mL/min; make-up gas: N_2 at 40 psi; operation mode: splitless (electronic pressure control); purge off time: 2 min; injection volume: 1 µL.

For PAHs analysis, a Hewlett- Packard 5890 gas chromatograph (GC) equipped with a Hewlett-Packard 5972 mass spectrometer (MS) was used. The gas chromatograph was installed with an HP-5 fused silica capillary column (30 m × 0.32 mm × 0.25 µm) (Hewlett- Packard, Germany). A 0.5 µL aliquot of the extract was manually injected with a syringe. The temperature program was initially set at 60°C (held at 60°C for 1 min) increased at a rate of 20°C/min to 300°C and then held at this temperature for 10 min. Helium was used as a carrier gas at a flow rate of 1.86 ml/min. For the mass spectrometer the following settings were used: ion source temperature and interface temperature 300°C. The analyses were operated using selected ion monitoring and electronic ionization was used for this purpose.

Table 2 and 3 show the quality parameters of the methods for OCPs and PAHs analysis.

2.5.4. GC-ECD and GC-MS Methods Analytical Performance

Since plant matrices with certified concentrations of pesticides are not available, fortified ecological plants (representative matrices) were analyzed during the validation study to verify the recovery, linearity, precision, limit of detection and limit of quantification. Validation was made in agreement with quality criteria described in document SANCO No 3131/2007 [24] concerning the pesticide and PAHs residues in food and feed. Representatives matrices were spiked, prior the extraction step. In this purpose solutions containing the pesticides and the PAHs respectively, at five levels of concentrations were used.

Specificity and selectivity are measures that assess the reliability of measurements in the presence of interferences.

For the repeatability conditions, the quantification of the same representatives matrices spiked with a solution containing the pesticides, respectively the PAHs at 1 mg/L level of concentration was performed ten times on the same day.

Table 2. Quality parameters of the method for pesticides determination

No	Compound	Retention time (min)	LOD* (μg /Kg)	LOQ (μg /Kg)
1	Lindane	11.15	0.11	0.3
2	p,p'-DDT	16.98	0.12	0.2
3	p,p'-DDE	15.78	0.12	0.2
4	p,p'-DDD	16.22	0.12	0.2
5	HCB	10.53	0.08	0.3
6	Aldrin	13.42	0.08	0.2
7	Dieldrin	15,78	0.11	0.2
8	Endrin	15,97	0.11	0.3
9	Heptachlor	11,35	0.08	0.2

Precision has been assessed on the basis of the relative standard deviation calculated from results generated under repeatability (RSD) conditions. The RSD values were less than 4.7% for both methods, indicating that RSD are below those given by the Horwitz equation. These results showed that the CG-ECD and GC-MS methods are precise.

Linearity was assessed on the base of the coefficient of determination (R^2), calculated by linear regression after having plotted the targeted spiking levels against the mean introduced concentrations.

Table 3. Quality parameters of the method for PAHs determination

No	Compound	Retention time (min)	m/z	LOD (ng /g)	LOQ (ng /g)
1	Np	5.74	128	0.0180	0.1
2	Acy	7.90	152	0.0020	0.1
3	Ace	8.13	154	0.0020	0.1
4	F	8.84	166	0.0018	0.1
5	Ph	10.07	178	0.0020	0.1
6	An	10.12	178	0.0040	0.2
7	Fl	11.66	202	0.0018	0.1
8	Py	11.94	202	0.0010	0.1
9	B[α]An	13.51	228	0.0020	0.1
10	Chry	13.76	228	0.0010	0.1
11	B[k]Fl	15.39	252	0.0020	0.1
12	B[α]Py	15.98	252	0.0010	0.1
13	B[ghi]P	18.87	276	0.0010	0.1
14	dB[α,h]An	18.98	278	0.0020	0.1
15	I[1,2,3-cd]Py	19.68	276	0.0010	0.1

The calculated R^2 values were above 0.998 for both methods, indicating the linear relationship between targeted spiking levels and mean introduced concentrations, within the working range of concentrations. The working concentration ranges were established by analyzing (n=10) each of two concentrations (the lowest and the highest concentration values of proposed concentration range). The homogeneity variance test was applied for these values. Just as for FAAS, calculated P values were compared with F_{tab} value (5.35 for n-1=9 free degrees) and obtained results showed that the calculated P values was below the F_{tab} value meaning that for our working ranges no significant differences were found between the variances of the concentration range limits. So, that the working ranges were correctly choosen.

LOD and LOQ values were determined using calibration standards. LOD and LOQ were calculated as $(3 \cdot S_a \text{-} a)/b$ and $(10 \cdot S_a \text{-} a)/b$, respectively, where b is the slope of the calibration curve and S_a is the standard deviation of intercept of regression equation. The LOD and LOQ values are presented in tables 2 and 3.

The accuracy of the method was further assessed by recovery studies (standard addition method), each test being performed six times. Satisfactory results were found, with recoveries between 96 and 99% indicating the high accuracy of the proposed method. These percentage recoveries were ranged between the limits imposed by Horwitz equation (85-110%) for the established concentration range. These results reveal that OCPs and PAHs could be accurately determined by CG-ECD respectively GC-MS methods.

3. RESULTS AND DISCUSSION

3.1. Statistical Analysis

To determine if the differences between samples from urban and rural areas were significant, among the different sources, a t-test was performed. To verify if there is variability between these samples an analysis of variance (ANOVA) was carried out. All statistical analyses were carried out at the 95% confidence level. According to this test, statistically significant differences were found between samples from urban and rural areas, respectively, among samples within each botanical origins (see table 4).

3.2. Heavy Metals

It can be notice that the concentration of metals in the plant varies with the stage development and studied metals exhibit a different distribution pattern. In both areas, urban (Constanta) and rural (Slava Rusa) in root of 5 cm tomato plant we found the highest concentration for Fe and this suggest us that the young plant absorb a higher quantity of Fe from soil. When plant grows at 35 cm we observe that Fe was found in higher quantity in leaves not in root.

Data regarding levels of heavy metals in tomato and bell pepper plants at different stage of growing (5, 15 and 35 cm) show that Cd and Cu have highest values in urban area than in rural area [14]. These values registering for plant's leaves indicate the atmospheric pollution.

Several authors have shown a relationship between atmospheric depositions and elevated element levels in crops, vegetables and soils. The contamination of soil by atmospheric deposition of toxic metals affects soil properties and further increase plant metal levels through root uptake [25]. Moseholm et al. [26] observed a linear relationship between air-borne Pb and its foliar concentrations in kale and Italian rye grass and showed that the magnitude of uptake was dependent upon atmospheric concentrations. Regarding Cu, soil appeared almost equally responsible for raised levels in leaves. This might be due to mobilization of Cu to leaves after uptake from soil.

Table 4. ANOVA design for pollutants content

Pollutants	Variation sources	d.f.	Sum of squares	F	p value
Heavy metals	Between rural and urban areas	2	583.8	15.83	0.000
	Between botanical origins	2	12,18	7.2	0.009
OCPs	Between rural and urban areas	2	2.88	4.72	0.031
	Between botanical origins	2	12.40	4.13	0.043
PAHs	Between rural and urban areas	2	0.9610	23.81	0.000
	Between botanical origins	2	0.5452	4.78	0.017

The highest Cd concentration was found for 35 cm bell pepper plant's leaves (0.159 mg/Kg), the highest Cu concentration was found for 35 cm tomato plant's leaves (6.98 mg/Kg) while in 5cm tomato plant's leaves was found the highest concentration for Zn (27.25 mg/Kg). All these values were obtained in urban area. Fe concentrations were in the normal limits for studied plants while Pb concentrations were under quantification limits.

In Bulgaria, Ivanova et al.[27] found in tomatoes leaves 4.210 mg/Kg of Cu levels, results that are in concordance with our levels of Cu. Al-Lahham et al. [28] were investigate the extent of translocation of heavy metals to tomato fruit produced in an open field near to Abu-Nusiar Wastewater Treatment Plant, Amman-Jordan. Seedlings were planted during the seasons of 1999 and 2000 and furrow irrigated with different mixtures of potable water to treated wastewater. The obtained results revealed an increase in the concentrations of Cu, Mn and Fe, in the soil correlated with high concentrations found in the wastewater and an increased concentration of Fe, Cu, Ni, Mn and Zn in the tomato and no accumulation of Cd and Pb. Kakar et al. [29] have investigated the effect of sewage waste water irrigation on heavy metal contents in the soil and tomato grown in Pakistan. The concentration of the Fe, Mn, Zn, Cu, Pb was 134.95, 108.20, 170.24, 21.13, 5.30, ppm in soil and 642.81, 475.20, 355.84, 30.91, 11.09 ppm in tomatoes, respectively. The findings of the study further demonstrated that the concentration of heavy metals was higher in tomatoes than soil. In our study Cu and Mn were found in smaller quantity than in Kakar's study. Gergen et al. [30] were found metals concentration from peppers cultivated in Banat area from Romania in concordance with our study.

Regarding metals concentration from cherry, sour cherry and apricot [17], the highest Cd concentration was found in almost ripe sour cherry (0.647 mg/Kg) while Pb was not detected in studied fruits. Saracoglu et al. [31] have found in dried apricots the following concentrations: Cu 0.92-6.49 ppm; Cd 0.02-0.72 ppm and Pb 0.72-3.77 ppm. These results are in concordance for Cu, Cd and higher for Pb than our results.

In nectarines, peaches, quinces and apples Cd and Pb concentrations were under the quantification limits. The highest concentrations for Cu and Zn were found in peaches ripe fruits (4.35 mg/Kg, respectively 11.05 mg/Kg) from rural area (Mereni). Poston [32] studied the content of metals in quince fruit and leaves and found the following levels: Cd (0.053-0.079)mg/kg, Cr (0.2) mg/kg and Pb (0.036-0.11) mg/kg. In our case all three elements were found in higher quantities than these results show in literature for quince.

Heavy metal concentration varied with species and with stage of growing of fruit considered for analysis. This is due probably to variable capabilities of plants to absorb and accumulate heavy metals. Furthermore, variations in growth period and growth rates as well as in physical and chemical properties of soil also affect heavy metal uptake.

3.2.1. Potential Impacts of Recorded Heavy Metals on Consumers' Health

According to Order no. 975/1998 issued by the Romanian Ministry of Public Health [33] the maximum limits for heavy metals in fruits are: are 0.05 mg/Kg for Cd, 0.5 mg/Kg for Pb and 5 mg/Kg for Cu and Zn.

Concentrations of Cd found in tomato plants from Constanta town are situated over the limits (0.138-0.149mg/kg) imposed by the Romanian Ministry of Public Health while for bell pepper Cd concentrations are exceeded just for 5 cm plant's leaves (0.15 mg/Kg), for stem (0.111 mg/Kg) and 35 cm plant's leaves (0.159 mg/Kg). Cu is over the maximum limits in 5 and 35 cm plant's leaves from Constanta town.(6.23 respectively 6.98 mg/Kg). High levels of Cu in Constanta can be a result of the anthropogenic release of heavy metals into the environment through smelting, manufacturing, agriculture and waste disposal technologies.

It was noticed that all values of Cd concentrations determined for cherries and sour cherries from urban area (Cernavoda) are over the recommendable maximum limit. In general, the major pathways through which Cd is released to the environment are to the atmosphere during fuel combustion, smelting, incineration of trash; to water via sewage and wastewater discharging (water contamination during battery, paint, plastics production, during soldering) point and nonpoint source runoff associated with the application of phosphate fertilizers and biosolids (product of sewage sludge); and as leachate from soils, landfills, and groundwater. The release via incineration and smelting processes produces fine airborne particles that attach quickly to dry or wet particles that can then be transported relatively long distances from the source [17]. In Cernavoda area the major pathway through Cd is released is fuel combustion because the highway is very near the Cernavoda town.

Pb and Cu concentrations found in these fruits are lower than their maximum limit values.

Provisional tolerable weekly intake (PTWI) depends on the amount, consumption period and contamination level of consumed food. PTWI for Cd was established 0.007 mg Cd per kg body weight/week (FAO/WHO [34]) and 0.49 mg Cd per week for 70 kg person. The mean annual fruit consumption in Romania is 60 kg per person and year [35]. This is equivalent to 1250 g per person per week. The results were calculated for fruits. The minimum Cd content in fruits was found as 0.123 mg/kg in cherries while the maximum content was found in sour cherries (0.647mg/kg). The estimated PTWI ranges from 0. 1537 mg (1250 g x 0.123 mg/1000 g) to 0.8087 mg (1250 g x 0.647 mg/1000 g) per person for cadmium in fruits. In order not to exceed the established PTWI value, the maximum Cd concentration in fruit should be 0.392 mg/kg. Sour cherries fruit concentrations exceed this value. Plants can accumulate 10 times more Cd than the concentration in the soil. Consumption of food contaminated with Cd affects the liver and kidneys. Doses higher than 1 mg per kg body weight – which is equivalent to much more than the PTWI range estimated above - can cause changes in the circulatory system [36].

The Joint FAO/WHO Expert Committee on Food Additives established a PTWI for Cu of 3.5 mg/ kg body weight/week which was equivalent to 245 mg/week for a 70 kg adult. By using the means of weekly fruits consumption in Romania of 1250 g per person and minimum (0.7 mg/kg in green peaches) and maximum (4.85 mg/kg in sour cherries) Cu levels in fruits, the weekly intake calculated ranged from 0.875 mg (1250 g x 0.7 mg/1000 g) to 6.0625 mg (1250 g x 4.85 mg/1000 g) per person for Cu in studied fruits. As can be seen, the estimated PTWI of Cu in this study is far below the established PTWI.

Joint FAO/WHO Expert Committee on Food Additives established a PTWI for zinc of 7 mg/kg body weight/week which was equivalent to 490 mg/week for a 70 kg adult. By using the means of weekly fruits consumption in Romania of 1250 g per person and the minimum zinc content as 0.23 mg/kg in almost ripe quince and the maximum content in almost ripe cherries (8.71 mg/kg), the weekly intake calculated ranged from 0.2875 mg (1250 g x 0.23 mg/1000 g) to 10.8875 mg (1250 g x 8.71 mg/1000 g) per person for Zn in studied fruits. These results show that the estimated PTWI of Zn in this study is below the established PTWI.

3.3. Organochlorine Pesticides

DDD, DDE and DDT concentrations from studied vegetables and fruits were under quantification limits. In pepper plants from rural area (Slava Rusa) lindan (2.49 ppb) was found as highest concentration in 35 cm plant's root while in urban area (Constanta) the highest concentration was found for aldrin (4.6 ppb) in 5cm plant's roots. In tomato the highest concentrations were found for lindan in rural area and for heptachlor in urban area. For studied fruits dieldrin was also found in small quantities while aldrin was present in all samples. Pesticide concentrations in fruits varied between 0.254 ppb (aldrin in green quince) and 4.489 ppb (heptachlor in almost ripe sour cherry).

The uptake of semivolatile lipophilic compounds into plants has been addressed by numerous studies. High levels of contaminants in plant tissues have repeatedly been attributed to accumulation of chemicals from the atmosphere. Diffusion and advection are known to be the principle uptake processes of OCPs from soil to root surfaces, but this uptake is often limited by high affinity of these compounds to the soil organic matter and surface of roots. Translocation of OCPs from soil to the shoot tissues via transpiration stream is generally small for hydrophobic compounds. Re-volatilization of OCPs from soil with subsequent sorption to the leaves may occur, but has probably a low impact. The other option is a particle-facilitated transport when the soil particles are resuspended from the soil surface and subsequently deposited on the leaves [37, 38].

3.3.1. Potential Impacts of Recorded Organochlorine Pesticides on Consumers' Health

The maximum limits for the pesticides of interest admitted from European Commission are: lindan-0.01 mg/Kg, HCB-0.01 mg/Kg, heptachlor 0.01 mg/Kg, aldrin-0.01 mg/Kg, dieldrin-0.01 mg/Kg, endrin-0.01 mg/Kg, ΣDDT-

0.05 mg/Kg [39]. All OCPs recorded for studied vegetables and fruits were under the maximum limits imposed by European Commission.

Greater level of OCPs residues were found in food samples (especially strawberries) from Gaza Governorates, in Kuwait, certain fruits exceeded the maximum residue limits for pesticides specified by the monitoring agencies [40].

3.4. Polycyclic Aromatic Hydrocarbons

A very wide range of PAH concentration was observed in studied samples [41]. For peppers the highest value was obtain for dB[a,h]An (1.162 µg/kg) in root of 5 cm plant grown in urban garden while I[1,2,3-cd]Py was not detected in 35 cm plant and fruit grown in rural area. Except Np, we observed that, while plant is young (5 cm) PAHs concentrations were in higher quantities than in mature plants (35 cm). This can suggest the fact that the accumulation of PAHs can be due to soil pollution or water contaminated, because in root are higher quantities that in stem or leaves. Soil can be affected by atmospheric deposition and other sources such as river water and rain water.

In pepper's fruit grown in rural area, the highest concentration was found for Py (0.213 µg/kg), while I[1,2,3-cd]Py, B[ghi]P and dB[a,h]An, Ace, B[a]An, Chry and B[k]Fl were under quantification limit. In pepper's fruit grown in urban area, the highest concentration was found for Np (0.564 µg/kg), while Chry, B[k]Fl, B[ghi]P and dB[a,h]An were under quantification limit. Zohair [42], in Egipt, has found concentrations for peppers from 0.04 to 1.46 ppb, results that are in concordance with data presented in our study.

In 35 cm tomatoes plant grown in rural area the most of PAHs concentrations were under detection limit. In tomato fruit from urban area the highest concentration was found for An (0.671 µg/kg) and in tomato fruit from rural area the highest concentration was found for B[k]Fl (0.063 µg/kg). In our study levels of total PAHs for tomatoes indicated a smaller average than Camargo and Toledo [43] study (9.50 µg/kg in tomato).

In 5 cm root tomato plant grown in urban area, except I[1,2,3-cd]Py, all the rest of analyzed PAHs were under quantification limit. In 35 cm root tomato plant grown in urban area, except Np, Py, B[a]An and Chry that were under detection limit, the rest of analyzed PAHs concentrations are in higher quantities that in stem or leaves or fruit. So while plant is young (5 cm) PAHs concentrations are higher in stem or leaves and when plant is mature (35 cm) PAHs concentrations are higher in root. This suggest that in urban area, when plant is young foliar uptake is the primary transfer pathway of PAHs from environment to vegetable and then when plant became mature PAHs concentrations are accumulated in root.

In rural area, when tomato plant is young (5 cm), except B[k]Fl, B[a]Py and dB[a,h]An (only in stem), PAHs concentrations were higher than in 35 cm plant. Stem and leaves have higher quantities that root system. So, PAHs in plant tissue may originate from volatile compound absorption by leaves in the surrounding air, deposition of contaminated soil particles, and dust on leaves, followed by retention in cuticle or penetration through it, and soil to root transfer followed by subsequent translocation by the transpiration stream [44]. The most PAHs concentrations from peppers and tomatoes grown in urban area were higher than those from peppers and tomatoes grown in rural area. The principal factor was maybe the city pollution, so PAHs levels in the atmosphere directly influenced PAHs levels in vegetable. There are many sources of PAHs in the atmosphere and emitted PAHs should be mixed during their passage through the atmosphere away from the source regions to other locations where the vegetables was collected [45]. In studied fruits, majority of PAH concentrations were found under limits of quantification while some of them were in the range 0.1009 ppb (B[a]An in almost ripe cherry) - 0.957ppb (Np in green quince).

3.4.1. Potential Impacts of Recorded Polycyclic Aromatic Hydrocarbons on Consumers' Health

BaP concentration is a good marker of carcinogenic PAH levels in environmental samples. BaP is the most known and studied member of the PAH because it is one of the most potent PAH animal carcinogens. BaP is found in coal tar, in automobile exhaust fumes (especially from diesel engines), tobacco smoke, and in charbroiled food [46, 47]. Maximum limit for BaP established by the Council Directive (1988) [48] was 0.03 µg/kg for foodstuffs. Some levels of BaP for studied vegetables and fruits were over the maximum limit.

CONCLUSION

The present work provides data regarding levels of heavy metals, OCPs and PAHs in flowering plants from Solanaceae and Rosaceae families. Studied plants at various growing stages were: tomato and bell pepper plants at 5, 15 and 35 cm (samples were taken from roots, stems, leaves and fruits) and peach, nectarine, apricot, cherry, sour cherry, apple and quince trees (green, almost ripe and ripe fruits) from Romania's urban and rural areas.

The concentrations of studied pollutants were not especially high in comparison to levels reported from fruits and vegetables reported in the literature, but their presence indicates a significant degree of pollution and permits the identification of principal contamination sources. In conclusion, the presence of these pollutants in flowering plants from Solanaceae and Rosaceae families raises some serious questions concerning their potential impacts on consumers' health, being necessary to monitor their distribution in the environment.

REFERENCES

[1] Pojana G., Critto A., Micheletti C., Carlon C., Busetti F., Marcomini A. (2003), Analytical and Environmental Chemistry in the Framework of Risk Assessment and Management: The Lagoon of Venice as a Case Study, *CHIMIA*, *57*(9), 542–549.

[2] Soceanu A., Dobrinas S., Coatu V., Chirila E. (2009), Pesticide residues determination in vegetables from Romania by GC-ECD, *Exposure And Risk Assessment Of Chemical Pollution – Contemporary Methodology. Springer Science-Business Media B.V.,* 455-462.

[3] Gavrilescu M. (2004), Removal of heavy metals from the environment by biosorption, *Engineering in Life Sciences*, *4*, 219-232.

[4] Sharma R.K., Agrawal M., Marshall F.M. (2008), Heavy metals (Cu, Cd, Zn and Pb) contamination of vegetables in Urban India: A Case Study in Varanasi, *Environ Poll*, *154*, 254 – 263.

[5] Ona L. F., Alberto A. M., Prudente J. A., & Sigua G. C. (2006), Levels of lead in urban soils from selected cities in a central region of the Philippines, *Environmental Science and Pollution Research, 13(3)*, 177–183.

[6] Deckert J. (2005), Cadmium toxicity in plants: Is there any analogy to its carcinogenic effect in mammalian cells? *BioMetals, 18(5),* 475–481.

[7] Oymak T., Tokalioglu S., Yılmaz V., Kartal S., Aydin D., (2009), Determination of lead and cadmium in food samples by the coprecipitation method, *Food Chemistry 113,* 1314–1317.

Risk Assessment of Inorganic and Organic Pollutants ...   67

[8]   Soceanu A., (2010), Determination of some trace metal concentrations in imported fruits by F-AAS and ICP-MS, *Environmental Engineering and Management Journal, 9(8)*, 1039-1044.
[9]   Abhilash P.C., Singh N. (2009), Pesticide use and application: An Indian scenario, *Journal of Hazardous Materials, 165, 1-12.*
[10] FAO/UNEP/OECD/SIB, (2001), Baseline Study on the Problem of Obsolete Pesticides Stocks, Rome.
[11] Betianu C., Gavrilescu M. (2006), Persistent organic pollutants in environment: inventory procedures and management in the context of the Stockholm convention, *Environmental Engineering and Management Journal*, 5(5), 1011-1028.
[12] Condurateanu S., Cadariu A., (2006), International project for POP-IPEP elimination, http://eea.ngo.ro/materiale/Tufarisuri_rom_final.pdf
[13] Nadal M., (2005), Human health risk assessment of exposure to environmental pollutants in the chemical/ petrochemical industrial area of Tarragona (Catalonia, Spain), European PhD thesis.
[14] Soceanu A., Dobrinas S., Popescu V., Magearu V.(2007), Determination of some trace metals in different stages of *Solanum Lycopersicum* plant growing, *Ovidius University Annals of Chemistry, 18(2)*, 120-123.
[15] Encyclopædia Britannica, http://www.britannica.com/EBchecked/topic/552838/Solanaceae.
[16] http://science.jrank.org/pages/5925/Rose-Family-Rosaceae.html
[17] Soceanu A. (2009), Presence of heavy metals in fruits from *Prunus* genera, *Ovidius University Annals of Chemistry, 20(1)*, 108-110
[18] Soceanu A., Dobrinas S., Popescu V., Birghila S., Magearu V. (2006), Selected heavy metals in fruits from Rosaceae Family, *Ovidius University Annals of Chemistry 17*, 1-4.
[19] EPA methods 8270C and 3600C.
[20] Ceausescu D., (1982), *Use of Mathematical Statistics in Analytical Chemistry*, (in Romanian), Technical Publishing House, Bucharest, Romania.
[21] Tanase I.G., Pana Al., Radu G.L., Buleandra M., (2007), *Validation of Analytical Methods. Theoretical Principles and Case Study*, Printech Press, Bucharest, Romania.
[22] Gonzalez A.G., Herrador A., (2007), A practical guide to analytical method validation, including measurement uncertainty and accuracy profiles, *Trends in analytical chemistry, 26(3)*, 227-238.
[23] Horwitz W., (1982), Evaluation of analytical methods used for regulation of foods and drugs, *Analytical Chemistry, 54*, 67A–76A.

[24] Method validation and quality control procedures for pesticide residues analysis in food and feed. Document SANCO/2007/3131.

[25] Pandey J., Pandey U., (2009), Accumulation of heavy metals in dietary vegetables and cultivated soil horizon in organic farming system in relation to atmospheric deposition in a seasonally dry tropical region of India, *Environ Monit Assess, 148*, 61–74.

[26] Moseholm L., Larsen E. H., Andersen B. and Nielsen M. M. (1992), Atmospheric deposition of trace elements around point sources and human health risk assessment. I: Impact zones near a source of lead emissions, *The Science of theTotal Environment 126*, 243-246.

[27] Ivanova J., Korhammer S., Djingova R., Heidenreich H., Markert B. (2001), Determination of lanthanoids and some heavy and toxic elements in plant certified reference materials by inductively coupled plasma mass spectrometry, *Spectrochimica Acta., 56,* 3-12.

[28] Al-Lahham O., El Assi N.M., Fayyad M., (2007), Translocation of heavy metals to tomato (Solanum lycopersicom L.) fruit irrigated with treated wastewater, *Scientia Horticulturae*, 113 (3), 250-254.

[29] Kakar R.G., Kakar K.M., Huq I., Huq A., Nasar M.H., Kakar S.R., (2005), Indus. Journal of Biological Sciences, *2(3),* 300

[30] Gergen I., Gogoasa I., Dragan S. and Moigradean D., Harmanescu M., (2006), Proc. of 7th Int. Symp. of Romanian Academy-Branch, Timisoara

[31] Saracoglu S., Tuzen M. and Soylak M. (2009), Evaluation of Trace Element Contents of Dried Apricot Samples from Turkey, *Journal of Hazardous Materials*, 167, 647-652.

[32] Poston T.M., Soil and Vegetation Surveillance, (1997), Annual Env. Report.

[33] Order 975, (1998), Order no. 975/1998 of Romanian Ministry of Public Health on the maximum permissible limits of metals in food, Romanian Official Monitor 268/11 juke 1999.

[34] FAO/WHO Expert Committee on Food Additives, http://www.who.int/ipcs/food/jecfa/en/.

[35] National Strategy (2008), National Strategy for Operational Programs for Fruits and Vegetables Sector, Romania. On line at: http://www.madr.ro/pages/vegetal/strategia-nationala-fructe-legume.pdf.

[36] Costin G.M., (2008), *Ecological Food*, (in Romanian), Romanian Academy Publishing House, Bucharest, Romania.

[37] Mikes O., Cupr P., Trapp S., Klanova J (2009), Uptake of polychlorinated biphenyls and organochlorine pesticides from soil and

air into radishes (Raphanus sativus), *Environmental Pollution 157*, 488–496.

[38] Soceanu A., Dobrinas S., Popescu V., Birghila S., Coatu V., Magearu V. (2006), Determination of some organochlorine pesticides and heavy metals from *Capsicum Annuum L.*, *Environmental Engineering and Management Journal, 5(4),* 597-604.

[39] Oficial Journal of European Union-Commission Regulation (EC) no. 149/2008 amending Regulation (EC) no. 396/2005 of the European Parliament and of the Council by establishing Annexes II, III and IV setting maximum residue levels for products.

[40] Abdulrahman O. Musaiger, Jassim H. Al-Jedah, Reshma D'souza (2008), Occurrence of contaminants in foods commonly consumed in Bahrain, *Food Control 19,* 854–861.

[41] Soceanu A., Dobrinas S., Stanciu G., Coatu V., Epure D.T. (2009), Residues of polycyclic aromatic hydrocarbons in different stages of *Capsicum annuum* (pepper) and *Solanum lycopersicum* (tomato) growing, *Environmental Engineering and Management Journal, 8(1),* 49-54

[42] Zohair A., (2006), Levels of polyaromatic hydrocarbons in Egyptian vegetables and their behavior during soaking in oxidizing agent solutions, *World J. of Agric. Sci., 2,* 90-94.

[43] Rojo Camargo M.C., Toledo M.C.F., (2003), Polycyclic aromatic hydrocarbons in Brazilian vegetables and fruits, *Food Control, 14,* 49-53.

[44] Fismes J., Perrin-Ganier C., Empereur-Bissonnet P., Morel J.L., (2002), Soil-to-root transfer and translocation of polycyclic aromatic hydrocarbons by vegetables grown on industrial contaminated soils, *J. Environ. Qual., 31,* 1649-1656.

[45] Liu X., Korenaga T., (2001), Dynamics analysis for the distribution of polycyclic aromatic hydrocarbons in rice, *J. of Health Science, 47,* 446-451.

[46] Kazerouni N., Sinha R., Hsu C-H., Greenberg A., Rothman N., (2001), Analysis of 200 food items for benzo[a]pyrene and estimation of its intake in an epidemiologic study, *Food and Chemical Toxicology, 39,* 423-436.

[47] http://www.answers.com/topic/benzopyrene

[48] Council Directive, (1988), Council Directive 88/388/EEC of 22 June 1988 on the approximation of the laws of the Member States relating to

flavourings for use in foodstuffs and to source materials for their production, OJ L 184, 15.7.1988, 61.

In: Flowering Plants
Editor: Jeremy J. Tellstone

ISBN: 978-1-61324-653-5
© 2011 Nova Science Publishers, Inc.

Chapter 3

SEED GERMINATION AND THE SECONDARY METABOLITES, PLANT HORMONES

Mohammad Miransari[*]
Department of Soil Science, College of Agricultural Sciences,
Shahed University, Tehran, Qom Highway,
Tehran, 18151/159, Iran

ABSTRACT

Seed germination is interesting and complicated. It is controlled by different mechanisms and is necessary for the growth and development of embryo, which eventually produces a new plant. Under unfavorable conditions seeds become dormant to maintain their germination ability. However, when the conditions are favorable seeds can germinate. There are different parameters controlling seed germination and dormancy, among which plant hormones are the most important ones. Plant hormones are produced by both plants and soil bacteria and control different processes related to seed growth and development. In this review article some of the most recent findings regarding seed germination and dormancy as well as the significance of plant hormones including abscisic acid, ethylene, gibberellins, auxin, cytokinins and brassinosteroids with reference to proteomic and molecular biology

---

[*]Correspondence information: Tel: (98)2151212243, Fax: (98)2151213192, E-mail: Miransari1@Gmail.Com, Miransari@Shahed.ac.ir

studies on such phenomena are discussed. In the conclusion some idea for the future research is expressed.

Keywords: Abscisic acid, auxin, brassinosteroids, cytokinins, ethylene, gibberellins, proteomic analysis, seed germination, soil bacteria.

INTRODUCTION

Plant hormones can significantly affect plants and microorganisms activities. It is speculated that plant hormones may have prokaryotic origin. Hence it can be interesting to evaluate the related effects such plant hormones can have on different plant activities including seed germination and dormancy. Plant hormones including abscisic acid (ABA), ethylene, gibberellins, auxin (IAA), cytokinins, and brassinosteroids are biochemical substances controlling many of physiological and biochemical processes in the plant. These interesting products are produced by plants and soil microbes (Finkelstein, 2004; Jimenez, 2005; Santner et al., 2009). There are hormone receptors with high affinity in the plant, responding to the hormones. For example, researchers have indicated that for ethylene and cytokinin the related receptors are the "sensory hybrid type histidine kinases", which are also active in the transduction pathways. It has been indicated that the receptors are of bacterial origin (Urao et al., 2000; Hwang and Sheen, 2001; Mount and Cheng, 2002; Santner et al., 2009).

Different methods have been used for the extraction of different biochemicals, including the plant and bacterial products, affecting the morphogenesis and physiological processes in the seed development. Such discoveries in combination with the use of exogenously used plant hormones (Lian et al., 2000) have made the advancement of the field more rapid and interesting (Miransari and Smith, 2009). Plant hormones are able to react interactively and hence the production of each may be dependent on the production of other hormones (Brady et al., 2003; Arteca and Arteca, 2008).

There are different parameters affecting the activities of plant hormones including the receptor properties like its affinity for the hormone, and cytosolic Ca^{2+}, which, for example, can influence stomatal activities through affecting the K^+ channel (Weyers and Paterson, 2001). Among their other functions, the effects of plant hormones on seed germination may be one of their most important effects on plant growth. Hence, in the following such effects with respect to proteomic and molecular biology studies are discussed with some prospects for future research.

SEED GERMINATION

Seed germination is a mechanism, in which different morphological and physiological alterations result in the germination of embryo. Before germination, seed absorbs water resulting in the expansion and elongation of seed embryo. When a part of embryo, the radicle, is grown out of the covering seed layers, the process of seed germination is completed (Hermann et al., 2007). Different researchers have evaluated the processes of seed germination, as affected by plant hormones in different plant families including Brassicaceae and Solanaceae (Muller et al., 2006; Hermann et al., 2007).

Seeds have proteins storage, like globulins and prolamins, which are increased during seed maturation, especially at the mid and late stage of seed maturation, when seeds absorb high amounts of nitrogen. The proteins are placed in the cell membrane or other parts of the seed. During the time of protein translocation into different parts of the seed, negligible amounts of protein are turned into other products. The activation of enzymes such as proteinase results in the mobilization of storage proteins (Wilson, 1986).

Storage proteins are also found in the seedling radicle and shoot (Tiedemann et al., 2000). The mobilization of proteins does not take place in different parts of the seed at the same time. The other enzymes, which become activated during the mobilization of proteins, are carboxypeptidase and aminopeptidase. Changes at molecular levels, which include the protein and hormonal alterations, and the balance between ABA and gibberellins, are among the most important parameters controlling the process of seed dormancy (Ali-Racheid et al., 2004; Finch-Savage and Leubner-Metzger, 2006; Finkelstein et al., 2008; Graeber et al., 2010).

Seed germination is influenced by different parameters including plant hormones. Use of mutants is one of the most interesting ways to indicate the role of each plant hormone in seed germination. The synthesis of DNA and mitotic microtubule are among the different changes taking place during embryogenesis and can be used as indicators of cell division and differentiation during embryogenesis. These processes are accompanied by seed abilities to tolerate desiccation and get dormant (Finkelstein, 2004; de Castro and Hilhorst, 2006).

Seed development includes the stage of embryo's body formation, which is related to cell division and differentiation resulting in the formation of embryonic organs (Goldberg et al., 1994, Meinke, 1995). This period is accompanied by maturation of seed including formation of organs and nutrient storage as well as changes in the embryo size and weight, followed by the acquirement of desiccation tolerance and dormancy (Finkelstein, 2004; de Castro and Hilhorst, 2006). Seed maturation results in the inhibition of cell cycle, decreased seed moisture, increased ABA level, production of storage reservoirs and established dormancy (Matilla and Matilla-Vazquez, 2008).

In addition to the effects of plant hormones on seed germination researchers have found that both under stress and non-stress conditions, N compounds including nitrous oxide can enhance seed germination through enhancing amylase activities (Zhang et al., 2005; Hu et al., 2007; Zheng et al., 2009). Through decreasing the production of O_2 and H_2O_2, such products can also alleviate the stress by controlling the likely oxidative damage, similar to the effects of antioxidant enzymes including superoxide dismutase (SOD), catalse (CAT) and peroxidase (POD) on plant growth under different stresses (Song et al., 2006; Tian and Lei, 2006; Tseng et al., 2007; Li et al., 2008; Zheng et al., 2009). In addition, N products can enhance seed germination by adjusting K^+/Na^+ ratio and increasing ATP production and seed respiration (Zheng et al., 2009). The allelopathic effects of seeds can also positively or adversely affect the germination of their own or other plant seeds (Ghahari and Miransari, 2009). Proteomic analysis of seed germination in *Arabidopsis taliana* indicated that during the process of seed germination 74 proteins are altered before radicle emergence and protrusion (Gallardo et al., 2001).

SEED DORMANCY

Seed dormancy is a mechanism by which seeds can inhibit their germination for more favorable conditions (Finkelstein et al., 2008). Usually freshly harvested seeds like barley (*Hordeum vulgare* L.) are not able to germinate at temperatures higher than $20^\circ C$ (Corbineau and Come, 1996; Leymarie et al., 2007). In Barley the process of dormancy is due to the fixation of oxygen by glumellae during the oxidation of phenolic products resulting in the limitation of oxygen supplement to the embryo. The resulted hypoxia may also interfere with ABA activities in the seed (Benech-Arnold et al., 2006). Gibberellins are able to activate the dormant seeds, although the hormone does not control seed dormancy (Bewley, 1997). ABA can inhibit corn germination

Seed Germination and the Secondary Metabolites, Plant Hormones   75

through affecting cell cycle. This is the reason for the more rapid germination of seeds, which are deficient in ABA. The inhibition of the cell cycle by ABA is related to the activation of a residual G1 kinase, which becomes inactivated in the absence of ABA (Sanchez et al., 2005).

By affecting hormonal balance in the seed, environmental parameters including salinity, acidity, temperature and light, can influence seed germination (Ali-Rachedi et al., 2004; Alboresi et al., 2006). Nitrate (NO_3^-) and gibberellins are able to enhance seed germination. NO_3^- can act as a source of N and a seed germination enhancer. Similarly, gibberellins enhance seed germination by inhibiting ABA activity. It is resulted by the activation of catabolyzing enzymes and inhibition of the related biosynthesis pathways, which also decreases ABA amounts (Toyomasu et al., 1994; Atia et al., 2009). Enzymes including nitrite reductase, nitrate reductase, and glutamine synthetase assimilate NO_3^- into amino acids and proteins.

Salinity decreases seed germination by affecting the seed nitrogen (N) content and hence embryo growth. Accordingly, this indicates how N compounds can alleviate the stress of salinity on seed germination (Atia et al., 2009). N can also inhibit seed dormancy by decreasing the level of ABA in the seed (Ali-Rachedi, et al., 2004; Finkelstein et al., 2008). The unfavorable effects of salinity on seed germination include: 1) decreasing the amounts of seed enhancer products including NO_3^- and gibberellins, 2) enhancing ABA amounts, 3) altering membrane permeability and water behavior in the seed (Khan and Ungar, 2002).

ABA and gibberellins are necessary for dormancy initiation and seed germination, respectively (Groot et al., 1992; Matilla and Matilla-Vazquez, 2008). The gibberellins/ABA balance determines seed ability to germinate or the pathways necessary for seed maturation (White et al., 2000; White and Rivin, 2000; Chibani et al., 2006; Finch-Savage and Leubner-Metzger, 2006). While ABA determines seed dormancy and inhibits seed from germination, gibberellins are necessary for seed germination (Matilla and Matilla-Vazquez, 2008).

Although seed dormancy is under the influence of plant hormones, seed morphological and structural characters like endosperm, pericarp and seed coat properties can also affect seed dormancy (Kucera et al., 2005). Both ethylene and gibberellins affect radicle growth with gibberellins being the most important hormone. Although gibberellins are necessary for the production of mannanase, which is necessary for seed germination, ethylene is not (Wang et al., 2005a). However, in gibberellins deficient mutants, ethylene can act similar to gibberellins, because the seeds are able to germinate completely in such a situation (Karssen et al., 1989; Matilla and Matilla-Vazquez, 2008).

Using proteomic analyses the molecular and biological stages related to seed germination have been indicated. At different stages of seed germination, expression of different genes results in the production of proteins, necessary for seed germination and dormancy release. Proteins necessary for seed germination are accumulated after-ripening, under seed dry conditions, resulting in the release of dormancy (Gallardo et al., 2001; Chibani et al., 2006).

ABSCISIC ACID (ABA)

While ABA positively affects stomatal activities, seed dormancy and plant activities under stresses like flooding and pathogens presence (Morre, 1989; Davies and Jones, 1991; Weyers and Paterson, 2001; Popko et al. 2010), it adversely affects the process of seed germination. For example, concentrations of 1-10 μM can inhibit seed germination in plants like *Arabidopsis taliana* (Kucera et al., 2005; Muller et al., 2006). However, other plant hormones including gibberellins, ethylene, cytokinins, and brassinosteroids as well as their negative interaction with ABA can positively regulate the process of seed germination (Kucera et al., 2005; Hermann et al., 2007). Under stress ABA can be quickly produced as a β-glucosidase hormone (Lee et al., 2006). Additionally, it has been indicated that phosphatase regulators can also act as ABA receptors (Ma et al., 2009).

The movement of ABA across the cellular membrane is under the influence of pH and cellular space. Hence, it is likely to predict the hormone concentration in different cellular compartments according to its cellular pH and spacing. Different experiments have specified that the receptors for ABA and IAA are located outside plasma membrane (Weyers and Paterson, 2001) indicating that sometimes appoplasm may be the important compartment. Soluble factors, F-boxes proteins, are receptors for IAA (Dharmasiri et al., 2005a, b; Kepinski and Leyser, 2005; Santner et al., 2009). For ABA one family of proteins called PYR/PYL/PAR is the receptor (Park et al., 2009). IAA is the most important hormone for the process of somatic embryogenesis (Cooke et al., 1993; Jimenez, 2005). Researchers have recently found two new G proteins, which are ABA receptors (Pandey et al., 2009).

The role of ABA and its responsive genes in the process of seed germination has been indicated (Nakashima et al., 2006; Graeber et al., 2010). The inhibitory effects of ABA on seed germination is through delaying the radicle expansion and weakening of endosperm as well as the enhanced transcription of factors, which may adversely affect the process of seed germination (Graeber et al., 2010). The gibberelin repressor RGL2 is able to inhibit seed germination by stimulating the production of ABA as well as the related transcription factors (Piskurewicz et al., 2008). The H subunit of a chloroplast protein, Mg-chelatase can act as ABA receptor during different growth stages of plant including seed germination (Shen et al., 2006). It has been recently suggested that GPROTEIN COUPLED RECEPTOR 2 can also act as another ABA receptor, mediating different activities of ABA including its effects on seed germination (Liu et al., 2007b). However, other researchers indicated that such a receptor is not a necessary for activities mediated by ABA including the process of seed germination (Johnston et al., 2007; Guo et al., 2008).

ETHYLENE

Compared with the other plant hormones, ethylene has the simplest biochemical structure. However, it can influence a wide range of plant activities (Arteca and Arteca, 2008). Similar to cytokinin, the perception of ethylene is by a kinase receptor, which is a two component protein. However, for ethylene the receptor is located in the membrane of endoplasmic reticulum (Kendrick and Chang, 2008; Santner et al., 2009). Although ethylene can affect different plant activities including tissue growth and development, and seed germination, however it is not yet indicated that how ethylene influences seed germination. There are different ideas regarding seed germination; according to some researchers ethylene is produced as a result of seed germination and according to the other researchers ethylene is necessary for the process of seed germination (Matilla, 2000; Petruzzelli et al., 2000; Rinaldi et al., 2000).

Ethylene is able to regulate plant responses, under different conditions including stresses. For example, in combination with ABA, ethylene is able to affect plant response to salinity. Under increased level of salinity, ethylene production in plant increases, which decreases plant growth and development. The enzyme 1-aminocyclopropane-1 carboxylic acid (ACC) is a pre-requisite for ethylene production, catalyzed by ACC oxidase. During the time that seed is exposed to the stress, ethylene production is affected (Mayak et al., 2004).

The amount of ethylene increases during the germination of different plant seeds like wheat, corn, soybean and rice affecting the rate of seed germination (Pennazio and Roggero, 1991; Zapata et al., 2004). ACC can enhance seed radicle emergence through the production of ethylene, produced in the radicle (Petruzzelli et al., 2000; 2003). With respect to the easy production of ethylene from ACC in the presence of ACC oxidase, ACC is widely tested in different experiments (Petruzzelli et al., 2000; Kucera et al., 2005).

It has been indicated that during the final stage of seed germination ethylene is produced in different plant species and it can contribute to the germination of seed after dormancy. Ethylene is produced through the pathway, which turns S-adenosyl-Met into 1-Amicocyclopropane-1-carboxillic-acid (ACC) by ACC synthase, followed by the oxidation of ACC to ethylene by ACC oxidase (Yang and Hoffman, 1984; Kende, 1993; Argueso et al., 2007). The amount of ethylene increases under stress and it can control different processes in the plant including flowering, fruit ripening, aging, dormancy inhibition and seed germination (Matilla, 2000; Nath et al., 2006; Matilla and Matilla-Vazquez, 2008).

As previously mentioned, ethylene is also important during stress and its production is affected, which influences plant growth (Druege, 2006). Researchers have indicated that ethylene in plant increases under stress, which can decrease plant growth including plant roots. They have also interestingly found that the bacterial enzyme ACC deaminase is able to alleviate such stresses by degrading the ethylene pre-requisite ACC (Mayak et al., 2004). Ethylene can also influence plant performance thorough affecting the production and functioning of other hormones for example by affecting the related pathways (Arora, 2005; Vandendussche and Van Der Straeten, 2007).

BR and IAA are able to stimulate the production of ethylene (Arteca and Arteca, 2008). Gibberellins, ethylene and BR can induce seed germination by rupturing testa and endosperm, while antagonistically interacting with the inhibitory effects of ABA on seed germination (Finch-Savage and Leubner-Metzger, 2006; Holdsworth et al., 2008; Finkelstein et al., 2008). Ethylene is able to make dormant seeds germinate. It has been indicated that through regulating the expression of cysteine-proteinase genes, and its complex protein, proteasome, ethylene can remove seed dormancy (Asano et al., 1999; Borgetti et al., 2002). These enzymes can degrade seed proteins during the first stages of germination.

Seed Germination and the Secondary Metabolites, Plant Hormones   79

The novel mechanism by which ethylene inhibits the adverse effects of ABA on the release of seed dormancy has been attributed to the production OH in the apoplasm. Production of reactive oxygen species in the apoplasm can also affect seed germination (Chen, 2008; Muller et al., 2009; Graeber et al., 2010). Reactive oxygen species are produced at different stages of seed growth and development. Usually reactive oxygen species can adversely affect seed activities. however, some new findings indicate that there are also some positive effects for reactive oxygen species including the germination of seeds and the growth of seedlings by the regulation of cell growth and development as well as controlling pathogens and cell redox conditions. Reactive oxygen species may also positively affect the release of seed dormancy through interactions with gibberellins and abscisic acid trusdcution pathways affecting so many transcriptional factors and proteins (El-Maarouf-Bouteau and Bailly, 2008).

GIBBERELLINS

Gibberellins belong to the family of tetracycline, diterpenoid, regulating plant growth. They are commonly used in modern agriculture and were first isolated from the metabolite products of the rice pathogenic fungus, *Gibberella fujikuroi,* in 1938 (Yamaguchi, 2008; Santner et al., 2009). The biosynthesis of gibberellins is from geranyldiphosphate through a pathway including several enzymes. Gibberellins are adversely regulated by DELLA proteins with C-terminal GRAS domain in their structure, which are eventually degraded by the hormone (Itoh et al., 2003; Schwechheimer, 2008).

Accumulation of DELLAs in seeds can result in the expression of genes producing F-box proteins. The receptor for gibberellins has been recently identified in rice. It is Gibberelline Insenstive Dawrf1 (GID1) protein (interacting with DELLA proteins and resulting in their eventual degradation) located in the nucleus and can bind to the gibberellins, which are biologically active (Ueguchi-Tanaka et al., 2005; Griffiths et al., 2006; Nakajima et al., 2006; Willige et al., 2007). Through its antagonistic effects with ABA, gibberellins, which are internal signals, are able to release dormancy in seeds (Gubler et al., 2008; Seo et al., 2009).

The seed endosperm, which becomes available to the embryo, by the activities of some hydrolases enzymes, is made of the starchy part and the surrounding aleurone (Jones and Jakobsen, 1991; Bosnes et al., 1992).

Gibberellins stimulate the synthesis and production of the hydrolases, especially α-amylase, resulting in the germination of seeds. Gibberellins are able to induce different genes, which are necessary for the production of amylases including α-amylase, proteases and β-glucanases (Appleford and Lenton, 1997; Yamaguchi, 2008). Different processes in the seed indicate that seed aleurone is appropriate for the evaluation of transduction pathways at the time of plant hormones production including gibberellins (Ritchie and Gilroy, 1998; Penfield et al., 2005; Achard et al., 2008; Schwechheimer, 2008).

The plant hormone gibberellins are necessary for seed germination. They can stimulate seed germination by different pathways including "weakening of endosperm" and "expansion of embryo cell". Proteins resulting in the modification of cell wall like xyloglucan endotransglycosylase/hydrolases (XTHs) and expansins may enhance the above mentioned pathways (Liu et al., 2005). The interesting mechanism, which controls seed germination, is the suppressing effects of excess ABA on embryo expansion, which inhibit the promoting effects of gibberellins on radicle growth and hence it will not germinate through the endosperm and testa (Nonogaki, 2008).

IAA

Auxin is a plant hormone, which is necessary for the following functions: cell cycling, growth and development, formation of vesicular tissues (Davies, 1995), and pollen (Ni et al., 2002) and development of other plant parts (He et al., 2000). The growth and development of different plant parts including embryo, leaf and root is believed to be controlled by auxin transport (Liu et al., 1993; Xu and Ni, 1999; Rashotte et al., 2000; Benjamins and Scheres, 2008; Popko et al., 2010).

Auxin by itself is not a necessary hormone for seed germination. However, according to analysis of auxin related genes expression, auxin is present in the seed radicle tip during and after seed germination. In addition, it is necessary that microRNA60 inhibits auxin Response Factor10 during seed germination so that the seed can germinate. Such a controlling process is also necessary for the stages related to post-emergence growth including seed maturation. The mechanisms for such inhibitory effects have been attributed to the interactions with the ABA pathway (Liu et al., 2007a). Although IAA may not be necessary for seed germination, it is necessary for the growth of young seedlings (Bialek et al., 1992). The accumulated IAA in the seed cotyledon is the major source of IAA for the seedlings. In legumes, amide products are the major source of IAA in the mature seeds (Epstein et al., 1986; Bialek and Cohen, 1989).

The most important plant hormones for seed germination are ABA and gibberellines, which have inhibitory and stimulatory effects on seed germination, respectively. BR and ethylene have also enhancing effects on seed germination. Although IAA by itself may not be important for seed germination, its interactions and cross talk with gibberellins and ethylene may influence the processes of seed germination and establishment (Fu and Harberd, 2003; Chiwocha et al., 2005).

Alteration of auxin signaling pathway, by altering auxin response factor increases seed sensitivity to ABA, as mRNA60 may affect the ABA responsive genes through repressing auxin response factor (Liu et al., 2007b). Auxin can influence seed germination, when ABA is present (Brady et al., 2003). Accordingly, mRNA60 is able to regulate the cross-talk between IAA and ABA. However, the molecular mechanism related to the interactions and cross-talk between IAA and ABA is not known yet.

CYTOKININS

The cytokinins are derived from adenine molecules in which the N6 is replaced. Miller was the first to discover them in 1950's based on their ability to enhance cellular division in plant (Miller et al., 1955). Cytokinins are plant hormones, regulating different plant activities including seed germination. They are active in all stages of germination (Chiwocha et al., 2005; Nikolic et al., 2006; Riefler et al., 2006). They can also affect the activities of meristemic cells in roots and shoots, and leaf senescence. In addition, they are effective in nodule formation during N-fixing processes and other interactions between plants and microbes (Murray et al., 2007; To and Kieber, 2008; Santner et al., 2009). The production of active cytokinins is through the activity of a phosphoribohydrolase enzyme, turning the nucleotide into the free base (Santner et al., 2009).

Signaling in cytokinins is extremely similar to the two-component signaling in bacterial species (To and Kieber, 2008). In this kind of perception, the initiation of phosphorelay by ligand binding is related to kinases histidine and asparate. The perceivers, which are in nucleus, are able to phosphorylate the response proteins, which can negatively or positively regulate cytokinins signaling. Similar to auxin, cytokinins is also able to regulate so many genes including Cytokinin Response Factors (Rashotte et al., 2003; Santner et al., 2009).

Cytokinins are also able to enhance seed germination by the alleviation of different stresses such as salinity, drought, heavy metals and oxidative stress (Khan and Ungar, 1997; Atici et al., 2005; Nikolic et al., 2006). They can be inactivated by the enzyme cytokinin oxidase/dehydrogenase (Galuszka et al., 2001) catalyzing the cleavage of their unsaturated bond. Different activities of cytokinins such as their effects on seed germination have been attributed to the different functioning of cytokinins in different cell types (Werner et al., 2001).

Arabidopsis thaliana has three histidine kinases as receptors for cytokinins (Inoue et al., 2001; Yamada et al., 2001). Controlling seed size including embryo, endosperm and seed coat growth, is also among the functions of cytokinins. Endosperm and seed coat growth in Arabidopsis is followed by embryo growth, at the later stage of embryogenesis, which is not related to the final seed size (Mansfield and Bowman, 1993). There are some parameters controlling seed number with respect to the number of seeds and the most important of which is the available carbon source for seed utilization (Riefler et al., 2006).

BRASSINOSTEROIDS

Brassinosteroids (BR) are a new class of plant hormones, similar to the steroid hormones in other organisms (Rao et al., 2002; Bhardwaj et al., 2006; Arora et al., 2008). The cholestane hydroxylated derivatives produce BR and the C-17 side chain and rings and the related replacements determine the variations in the hormonal structure (Arora et al., 2008). BR have a wide range of activities in plant growth and development including cell growth, vascular formation, reproductive growth, seed germination, and production of flower and fruit (Khripach et al., 2000; Cao et al., 2005).

BR is able to enhance seed germination through controlling the inhibitory effects of ABA on seed germination (Finkelstein et al., 2008; Zhang et al., 2009). The perception of BR is through BRI1, a leucine receptor similar to kinase, located on the cell surface (Li and Chory, 1997). As a result of BR binding to BRI1 receptor, phosphorylation of sites, changes to the cytosolic domain and BKI1 dissociation (the receptor, adversely affecting BR signaling) in the plasma membrane take place (He et al., 2000; Wang et al., 2001; 2005b, Wang and Chory, 2006). Accordingly, the activation of BRI1 and its interaction with other kinase receptors or other substrates results in a set of reactions including the phosphorylation of some transcription factors in plant, indicating the level of hormone signaling (Yin et al., 2002; He et al., 2005; Wang and Chory, 2006). The increase in the rate of the phosphorylation indicates that ABA is able to inhibit BR activity through affecting the related genes (Zhang et al., 2009).

BR, gibberellic acid and ethylene are able to increase the growth of embryo out of seed by enhanced rupturing of endosperm and antagonistically interacting with ABA (Finch-Savage, Leubner-Metzger, 2006). These hormones are able to enhance seed germination by their own signaling pathway. While gibberellins and light are able to enhance seed germination through releasing seed photodormancy, BR can increase seed germination by enhancing the growth of embryo (Leubner-Metzger, 2001).

SOIL MICROORGANISMS AND PRODUCTION OF PLANT HORMONES

Similar to plants, soil microorganisms including plant growth promoting bacteria (PGPR) such as *Azospirillum* sp. and *Pseudomonas* sp. are also able to produce plant hormones as the secondary metabolites (Miransari, 2011). Such hormones are utilized as plant growth promoting substances at the time of inoculating the host plant (Johri, 2008; Abbas Zadeh et al., 2009; Jalili et al., 2009). These hormones include auxins, which is produced at the highest amount relative to the other plant hormones (Zimmer and Bothe, 1988), cytokinins (Cacciari et al., 1986) and gibberellins (Piccoli et al., 1996). It is speculated that production of plant hormones may have prokaryotic origin. It is because genes are exchanged between the two organisms as there are different microorganisms in the rhizosphere, which may result in the uptake of DNA by the plant (Bode and Muller, 2003).

It has been indicated that the production of IAA in the related pathway in *Azospirillum brasilence* is controlled by *ipdC* gene (Vande Broek et al., 1999; Spaepen et al., 2008), which is expressed in the stationary growth phase (Vande Broek et al., 2005). The gene *ipdC*, whose crystallographic structure has been recently indicated (Versees et al., 2007a; b), produces phenylpuruvate decarboxylase (Spaepen et al., 2007).

PGPR affect plant growth and soil properties through their activities including the production of plant hormones, different enzymes, siderophores, and HCN (Botelho and Mendonça-Hagler, 2006; Miransari, 2011) resulting in the enhanced plant growth and soil structure. For example, production of 1-amino-1-cyclopropane carboxylic acid (ACC) deminase by *Pseudomonas fluorescence* and *P. putida* can enhance plant growth under different stresses. ACC deminase is able to degrade ethylene, whose production increases under stress and adversely affects plant growth. In Addition, such activities also result in the enhanced nutrient availability and controlling pathogens (Lugtenberg et al., 1991; Nagarajkumar et al., 2004).

CONCLUSION

Seed germination and dormancy are important mechanisms affecting crop production. There are different factors including plant hormones. Plant hormones produced by plants and soil bacteria can significantly affect seed germination. The collection of plant hormones including ABA, IAA, cytokinins, ethylene, gibberellins and brassinosteroids can positively or adversely affect seed germination, while interacting with each other. The molecular pathways, recognized by proteomic and molecular biology analyses regarding the perception of plant hormones may indicate more details related to the effects of plant hormones on seed germination and dormancy. The important role of soil bacteria in the production of plant hormones and hence seed germination can be used as a very effective tool for the enhanced seed germination and hence crop production. Future research may focus on how the combination of appropriate agricultural strategies and biological methods like using soil bacteria can provide a proper medium for the germination and growth of seeds under different conditions.

REFERENCES

Abbas-Zadeh, P., Saleh-Rastin, N., Asadi-Rahmani, H., Khavazi, K., Soltani, A., Shoary-Nejati, A.R., Miransari, M. 2009. Plant growth promoting activities of fluorescent pseudomonads, isolated from the Iranian soils. *Acta Physiologiae Plantarum*, 32, 281-288.

Achard, P., Renou, J-P., Berthome, R., Harberd, N.P., Genschik, P. 2008. Plant DELLAs restrain growth and promote survival of adversity by reducing the levels of reactive oxygen species. *Current Biology*, 18, 656-660.

Ali-Rachedi, S., Bouinot, D., Wagner, M.H., Bonnet, M., Sotta, B., Grappin, P., Jullien, M. 2004. Changes in endogenous abscisic acid levels during dormancy release and maintenance of mature seeds: studies with the Cape Verde Islands ecotype, the dormant model of *Arabidopsis thaliana*, *Planta,* 219, 479–488.

Alboresi, A., Gestin, C., Leydecker, M.T., Bedu, M., Meyer, C, Truong, H.N. 2006. Nitrate, a signal relieving seed dormancy in *Arabidopsis, Plant Cell and Environment.* 28, 500–512.

Appleford, N.E.J., Lenton, J.R. 1997. Hormonal regulation of a-amylase gene expression in germinating wheat (*Triticum aestivum*) grains. *Physiologia Plantarum,* 100, 534-542.

Asano, M., Suzuki, S., Kawai, M., Miwa, T., Shibai, H. 1999. Characterization of novel cysteine proteinases from germinating cotyledons of soybean (*Glycine max* L. Merrill). *Journal of Biochemistry,* 126, 296-301.

Argueso, C.T., Hansen, M., Kieber, J.J. 2007. Regulation of ethylene biosynthesis. *Journal of Plant Growth Regulation,* 26, 92–105.

Arora, A. 2005. Ethylene receptors and molecular mechanism of ethylene sensitivity in plants, *Current Science,* 89,1348–1361.

Arora, N, Bhardwaj, R., Sharma, P., Arora, H. 2008. Effects of 28-homobrassinolide on growth, lipid peroxidation and antioxidative enzyme activities in seedlings of Zea mays L. under salinity stress. *Acta Physiologiae Plantarum,* 30, 833–839.

Arteca, R., Arteca, J. 2008. Effects of brassinosteroid, auxin, and cytokinin on ethylene production in Arabidopsis thaliana plants. *Journal of Experimental Botany,* 59, 3019-3026.

Atia, A., Debez, A., Barhoumi, Z., Smaoui, A. Abdelly, C. 2009. ABA, GA3, and nitrate may control seed germination of *Crithmum maritimum* (Apiaceae) under saline conditions. *Comptes Rendus Biologies,* 332, 704-710.

Atici, O., Agar, G., Battal, P. 2005. Changes in phytohormone contents in chickpea seeds germinating under lead or zink stress. *Biologia Plantarum,* 49, 215–222.

Benech-Arnold, R.L., Gualano, N., Leymarie, J., Come, D., Corbineau, F. 2006. Hypoxia interferes with ABA metabolism and increases ABA sensitivity in embryos of dormant barley grains. *Journal of Experimental Botany,* 57, 1423–1430.

Benjamins, R., Scheres, B. 2008. Auxin: the looping star in plant development. *Annual Review of Plant Biology,* 59, 443-465.

Bewley JD. 1997. Seed germination and dormancy. *The Plant Cell* 9, 1055–1066.

Bhardwaj, R., Arora, H.K., Nagar, P.K., Thukral, A.K. 2006. Brassinosteroids-a novel group of plant hormones. In: Trivedi PC (ed) *Plant molecular physiology-current scenario and future projections.* Aavishkar Publisher, Jaipur, pp 58–84.

Bialek, K., Cohen, J.D. 1989. Free and conjugated indole-3-acetic acid in developing bean seeds. *Plant Physiology,* 91, 398-400.

Bialek, K., Michalczuk, L., Cohen, J.D. 1992. Auxin biosynthesis during Seed germination in Phaseolus vulgaris. *Plant Physiology,* 100, 509-517.

Bode, H.B., Müller, R. 2003. Possibility of bacterial recruitment of plant genes associated with the biosynthesis of secondary metabolites. *Plant Physiology,* 132, 1153-1161.

Borghetti, F., Noda, F.N., de Sa C.M. 2002. Possible involvement of proteasome activity in ethylene-induced germination of dormant sunflower embryos. *Brazilian Journal of Plant Physiology,* 14, 125–131.

Bosnes, M., Weideman, F., Olsen, O.A. 1992. Endosperm differentiation in barley wild type and sex mutants. *Plant Journal* 2, 661-647.

Botelho, G.R., Mendonça-Hagler, L.C. 2006. Fluorescent Pseudomonads associated with the rhizospehre of crops- an overview. *Brazilian Journal of Microbiology,* 37, 401-416.

Brady, S.M., Sarkar, S.F., Bonetta, D. and McCourt, P. 2003. The ABSCISIC ACID INSENSITIVE 3 (ABI3) gene is modulated by farnesylation and is involved in auxin signaling and lateral root development in Arabidopsis. *Plant Journal,* 34, 67–75.

de Castro, R. D., Hilhorst, H.W.M. 2006. Plant Hormonal control of seed development in GA- and ABA-deficient tomato (*Lycopersicon esculentum* Mill. cv. Moneymaker) mutants. *Plant Science,* 170, 462–470.

Cacciari, I., Lippi, D., Pietrosanti, T., Pietrosanti, W. 1989. Phytohormone-like substances produced by single andmixed diazotrophic cultures of *Azospirillum* and *Arthrobacter. Plant and Soil,* 115, 151–153.

Cao, S., Xu, Q., Cao,Y.,Quian, K., An, K., Zhu,Y., Bineng, H., Zhao, H., Kuai, B. 2005. Loss of function mutations in DET2 gene lead to an enhanced resistance to oxidative stress in Arabidopsis. *Physiologia Plantarum,* 123, 57-66.

Chen, J. 2008. Heterotrimeric G-proteins in plant development. *Frontiers in Bioscience,* 13, 3321-3333.

Chibani, K., Ali-Rachedi, S., Job, C., Job, D., Jullien, M., Grappin, P. 2006. Proteomic analysis of seed dormancy in Arabidopsis. *Plant Physiology,* 142, 1493–1510.

Chiwocha, S.D., Cutler, A.J., Abrams, S.R., Ambrose, S.J., Yang, J., and et al. 2005. The etr1-2 mutation in *Arabidopsis thaliana* affects the abscisic acid, auxin, cytokinin and gibberellin metabolic pathways during maintenance of seed dormancy, moist-chilling and germination. *Plant Journal,* 42, 35–48.

Cooke, T.J., Racusen R.H., Cohen, J.D. 1993. The role of auxin in plant embryogenesis. *Plant Cell,* 5, 1494–1495.

Corbineau, F., Come, D. 1996. Barley seed dormancy. *Bios Boissons Conditionnement,* 261, 113–119.

Davies, W.J., Jones, H.G. 1991. *Abscisic acid: physiology and biochemistry.* Cambridge, UK: BIOS Scientific Publishers Ltd.

Davies PJ. *Plant Hormones,* Dordrecht. The Netherlands: Kluwer Academic Publishers, 1995.

Dharmasiri, N., Dharmasiri, S., Estelle, M. 2005a. The F-box protein TIR1 is an auxin receptor. *Nature,* 435, 441-445.

Dharmasiri, N., Dharmasiri, S., Weijers, D., Lechner, E., Yamada, M., Hobbie, L., Ehrismann, J.S., Jürgens, G., Estelle, M. 2005b. Plant development is regulated by a family of auxin receptor F box proteins. *Developmental Cell,* 9, 109–119.

Druege, U. 2006. Ethylene and plant responses to abiotic stress, in: N.A. Khan (Ed.), *Ethylene Action in Plants,* Springer-Verlag,Berlin, pp. 81–118.

El-Maarouf-Bouteau, H., Bailly, C. 2008. Oxidative signaling in seed germination and dormancy. *Plant Signaling and Behavior,* 3, 175-182.

Epstein, E., Baldi, B.G., Cohen, J.D. 1986. Identification of indole-3-acetyl glutamate from seeds of Glycine max L. *Plant Physiology,* 80, 256-258.

Finch-Savage W., Leubner-Metzger, G. 2006. Seed dormancy and the control of germination. *New Phytologist,* 171, 501–523.

Finkelstein, R.R. 2004. The role of hormones during seed development and germination, in: P.J. Davies (Ed.), *Plant Hormones: Biosynthesis, Signal transduction, Action!,* Kluwer Academic Publishers, Dordrecht, The Netherlands, pp. 513–537.

Finkelstein, R., Reeves, W., Ariizumi, T., Steber, C. 2008. Molecular aspects of seed dormancy. *Annual Review of Plant Biology,* 59, 387-415.

Fu, X., Harberd, N.P. 2003. Auxin promotes Arabidopsis root growth by modulating gibberellin response. *Nature,* 421, 740–743.

Gallardo, K., Job, C., Groot, S., Puype, M., Demol, H., Vandekerckhove, J., Job, D. 2001. Proteomic analysis of arabidopsis seed germination and priming. *Plant Physiology,* 126, 835-848.

Galuszka, P., Frebort, I., Sebela, M., Sauer, P., Jacobsen, S., et al. 2001. Cytokinin oxidase or dehydrogenase? Mechanism of cytokinin degradation in cereals. *European Journal of Biochemistry,* 268, 450–461.

Ghahari, S., Miransari, M. 2009. Allelopathic effects of rice cultivars on the growth parameters of different rice cultivars. *International Journal of Biological Chemistry,* 3, 56-70.

Goldberg, R.B., de Paiva, G., Yedegari. R. 1994. Plant embryogenesis: zygote to seed, *Science,* 266, 605–614.

Graeber, K., Linkies, A., Muller, K., Wunchova, A., Rott, A., Leubner-Metzger, G. 2010. Cross-species approaches to seed dormancy and germination: conservation and biodiversity of ABA-regulated mechanisms and the Brassicaceae DOG1 genes. *Plant Molecular Biology,* 73, 67–87.

Griffiths, J., Murase, K., Rieu, I., Zentella, R., Zhang, Z., Powers, S., Gong, F., Phillips, A., Hedden, P., Sun, T., Thomas. S. 2006. Genetic characterization and functional analysis of the GID1 gibberellin receptors in *Arabidopsis*. *Plant Cell*, 18, 3399-3414.

Groot, S.P.C., Karssen, C.M. 1992. Dormancy and germination of abscisic acid deficient tomato seeds: studies with the *sitiens* mutant. *Plant Physiology*, 99, 952–958.

Gubler, F., Hughes, T., Waterhouse, P., Jacobsen, J. 2008. Regulation of dormancy in barley by blue light and after-ripening: effects on abscisic acid and gibberellin metabolism. *Plant Physiolog,y* 147, 886–896.

Guo, J., Zeng, Q., Emami, M., Ellis, B., Chen, J. 2008. The GCR2 gene family is not required for ABA control of seed germination and early seedling development in Arabidopsis. *PLoS ONE*, 3, 1-7.

He, Y.K., Xue, W.X., Sun, Y.D., Yu, X.H., Liu, P.L. 2000. Leafy head formation of the progenies of transgenic plants of Chinese cabbage with exogenous auxin genes. *Cell Research*, 10, 151-602.

Hwang, I. and Sheen, J. 2001. Two-component circuitry in *Arabidopsis* cytokinin signal transduction. *Nature*, 413, 383-389.

Hermann, K., Meinhard, J., Dobrev, P., Linkies, A., Pesek, P., He, B., Machackova, I., Fischer, U., Leubner-Metzger, G. 2007. 1-Aminocyclopropane-1-carboxylic acid and abscisic acid during the germination of sugar beet (*Beta vulgaris* L.): a comparative study of fruits and seeds. *Journal of Experimental Botany*, 58, 3047–3060.

He, Z., Wang, Z.Y., Li, J., Zhu, Q., Lamb, C., Ronald, P., Chory, J. 2000. Perception of brassinosteroids by the extracellular domain of the receptor kinase BRI1. *Science*, 288, 2360–2363.

He, J., Gendron, J., Sun, Y., Gampala, S., Gendron, N., Sun, C., Wang, Z., 2005. BZR1 is a transcriptional repressor with dual roles in brassinosteroids homeostasis and growth responses. *Science*, 307, 1634–1638.

Holdsworth, M., Bentsink, L., Soppe, W. 2008. Molecular networks regulating Arabidopsis seed maturation, afterripening, dormancy and germination. *New Phytologist*, 179, 33–54.

Hu, K.D., Hu, L.Y., Li, Y.H., Zhang, F.Q., Zhang, H., 2007. Protective roles of nitric oxide on germination and antioxidant metabolism in wheat seeds under copper stress. *Plant Growth Regulation*, 53, 173–183.

Inoue, T., Higuchi, M., Hashimoto, Y., Seki, M., Kobayashi, M., Kato, T., Tabata, S., Shinozaki, K., and Kakimoto, T. 2001. Identification of CRE 1 as a cytokinin receptor from Arabidopsis. *Nature*, 409, 1060 - 1063.

Itoh, H., Matsuoka, M., Steber, C.M. 2003. A role for the ubiquitin-26S-proteasome pathway in gibberellin signaling. *Trends in Plant Science*, 8, 492-497.

Seed Germination and the Secondary Metabolites, Plant Hormones 89

Jalili, F., Khavazi, K., Pazira, E., Nejati, A., Asadi Rahmani, H., Rasuli Sadaghiani, H., Miransari, M. 2009. Isolation and characterization of ACC deaminase producing fluorescent pseudomonads, to alleviate salinity stress on canola (*Brassica napus* L.) growth. *Journal of Plant Physiology*, 166, 667-674.

Jimenez, V.M. 2005. Involvement of plant hormones and plant growth regulators on in vitro somatic embryogenesis. *Plant Growth Regulation*, 47, 91–110.

Johnston, C.A., Temple, B.R., Chen, J.G., Gao, Y., Moriyama, E.N., Jones, A.M., Siderovski, D.P., Willard, F.S. 2007. Comment on 'A G Protein-Coupled Receptor Is a Plasma Membrane Receptor for the Plant Hormone Abscisic Acid'. *Science*, 318, 914c.

Johri, M.M. 2008. Hormonal regulation in green plant lineage families. *Physiology and Molecular Biology of Plants*, 14, 23-38.

Jones, R.L., Jacobsen, J.V. 1991. Regulation of the synthesis and transport of secreted proteins in cereal aleurone. *International Reviews of Cytology*, 126, 49-88.

Karssen, C.M., Zaorski, S., Kepczynski, J., Groot. S.P.C. 1989. Key role for endogenous gibberellins in the control of seed germination, *Annals of Botany*, 63, 71–80.

Kende, H. 1993. Ethylene biosynthesis. Annual Reviews of Plant *Physiology and Plant Molecular Biology*, 44, 283–307.

Kendrick, M.D., Chang, C. 2008. Ethylene signaling: new levels of complexity and regulation. *Current Opinion in Plant Biology*, 11, 479-485.

Kepinski, S., Leyser, O. 2005. The Arabidopsis F-box protein TIR1 is an auxin receptor. *Nature*, 435, 446–451.

Khan, M.A., Ungar, I.A. 1997. Alleviation of seed dormancy in the desert forb *Zygophyllum simplex* L. from Pakistan. *Annals of Botany*, 80, 395–400.

Khan, M.A. Ungar, I.A. 2002. Role of dormancy relieving compounds and salinity on the germination of *Zygophyllum simplex* L., *Seed Science Technology* 30, 16–20.

Khripach, V., Zhabinskii, V., Groot, A.D. 2000. Twenty years of brassinosteroids: steroidal plant hormones warrant better crops for XXI century. *Annals of Botany*, 86, 441–447.

Kucera, B., Cohn, M.A., Leubner-Metzger, G. 2005. Plant hormone interactions during seed dormancy release and germination. *Seed Science Research*, 15, 281–307.

Leubner-Metzger, G. 2001. Brassinosteroids and gibberellins promote tobacco seed germination by distinct pathways. *Planta*, 213, 758-763.

Lee, K.H., Kim, H.-Y., Piao, H.L., Choi, S.M., Jiang, F., Hartung, W., Hwang, I., Kwak, J.M., Lee, I.-J., and Hwang, I. 2006. Activation of glucosidase via stress-induced polymerization rapidly increases active pools of abscisic acid. *Cell,* 126, 1109-1120.

Leymarie, J., Bruneaux, E., Gibot-Leclerc, S, Corbineau F. 2007. Identification of transcripts potentially involved in barley seed germination and dormancy using cDNA-AFLP. *Journal of Experimental Botany*, 58, 425–437.

Li, J., Chory, J. 1997. A putative leucine-rich repeat receptor kinase involved in brassinosteroid signal transduction. *Cell,* 90, 929–938.

Li, Q.Y., Niu, H.B., Yin, J.,Wang, M.B., Shao, H.B., Deng, D.Z., Chen, X.X., Ren, J.P., Li, Y.C., 2008. Protective role of exogenous nitric oxide against oxidative-stress induced by salt stress in barley (*Hordeum vulgare*). Colloid. Surface B: *Biointerface*, 65, 220–225.

Lian, B., Zhou, X., Miransari, M., Smith, D.L. 2000. Effects of salicylic acid on the development and root nodulation of soybean seedlings. *Journal of Agronomy and Crop Science*, 185, 187 – 192.

Liu, Chun-ming, Xu Zhi-hong, Chua, Nam-hai. 1993. Auxin polar transport is essential for the establishment of bilateral symmetry during early plant embryogenesis. *Plant Cell,* 5, 621-30.

Liu, P.P., Koizuka, N., Homrichhausen, T.M., Hewitt, J.R., Martin, R.C., Nonogaki, H. 2005. Large-scale screening of Arabidopsis enhancer-trap lines for seed germination-associated genes. *Plant Journal,* 41, 936-944.

Liu, P.P., Montgomery, T.A., Fahlgren, N., Kasschau, K.D., Nonogaki, H., Carrington, J.C. 2007a. Repression of AUXIN RESPONSE FACTOR10 by microRNA160 is critical for seed germination and post-germination stages. *Plant Journal*, 52, 133-46.

Liu, X., Yue, Y., Li, B., Nie, Y., Li, W., Wu, WH., Ma, L. 2007b. A G protein-coupled receptor is a plasma membrane receptor for the plant hormone abscisic acid. *Science,* 315, 1712–1716.

Lugtenberg, B.J.J., De Weger, A.L., Bennet, J.W. 1991. Microbial stimulation of plant growth and protection from disease. *Current Opinion in Biotechnology,* 2, 457-464.

Ma, Y, Szostkiewicz, I., Korte, A., Moes, D., Yang, Y., Christmann, A., Grill, E., 2009. Regulators of PP2C phosphatase activity function as abscisic acid sensors. *Science,* 324, 1064 - 1068.

Mansfield, S.G., Bowman, J. 1993. Embryogenesis. In *Arabidopsis: An Atlas of Morphology and Development*, J. Bowman, ed. Berlin: Springer-Verlag, pp. 349–362.

Matilla, A.J. 2000. Ethylene in seed formation and germination, *Seed Science Research,* 10, 111–126.

Matilla, A.J., Matilla-Vazquez, M.A. 2008. Involvement of ethylene in seed physiology. *Plant Science,* 175, 87–97.

Seed Germination and the Secondary Metabolites, Plant Hormones 91

Mayak, S., Tirosh, T., Glick, B.R. 2004. Plant growth-promoting bacteria confer resistance in tomato plants under salt stress. *Plant Physiology and Biochemistry*, 42, 565-572.

Meinke, D.W. 1995. Molecular genetics of plant embryogenesis. 1995. *Annual Reviews of Plant Physiology and Plant Molecular Biology*, 46, 369–394.

Miransari, M., Smith D.L. 2009. Rhizobial lipo-chitooligosaccharides and gibberellins enhance barley (Hoedum vulgare L.) seed germination. *Biotechnology*, 8, 270-275.

Miransari, M. 2011. Interactions between arbuscular mycorrhizal fungi and soil bacteria. Review article, *Applied Microbiology and Biotechnology*, 89, 917-930.

Miller, C., Skoog, F., Saltza, M.V., Strong, M. 1955. Kinetic, a cell division factor from deoxyribonucleic acid. *Journal of American Chemical Society* 77, 1392-1393.

Moore, T.C. 1989. *Biochemistry and Physiology of Plant Hormones*, 2nd edn. New York, USA: Springer-Verlag.

Mount, S.M., Cheng, C. 2002. Evidence for plastid origin of plant ethylene receptor genes. *Plant Physiology*, 130, 10-14.

Muller K., Tintelnot S., Leubner-Metzger, G. 2006. Endospermlimited Brassicaceae seed germination: abscisic acid inhibits embryo-induced endosperm weakening of *Lepidium sativum* (cress) and endosperm rupture of cress and *Arabidopsis thaliana*. *Plant and Cell Physiology*, 47, 864–877.

Muller, K., Linkies, A., Vreeburg, R.A.M., Fry, S.C., Krieger-Liszkay, A., Leubner-Metzger, G. 2009b. In vivo cell wall loosening by hydroxyl radicals during cress (Lepidium sativum L.) seed germination and elongation growth. *Plant Physiology*, 150, 1855-1865.

Murray, J., Karas, B., Sato, S., Tabata,S., Amyot,L., Szczyglowski, K. 2007. A cytokinin perception mutant colonized by Rhizobium in the absence of nodule organogenesis. *Science*, 315, 101-104.

Nagarajkumar, M., Bhaskaran, R., Velazhahan, R. 2004. Involvement of secondary metabolites and extracellular lytic enzymes produced by *Pseudomonas fluorescens* in inhibition of *Rhizoctonia solani*, the rice sheath blight pathogen. *Microbiological Research*, 159, 73-81.

Nakajima, M., Shimada, A., Takashi, Y., Kim, Y., Park, S., Tanaka, M., Suzuki, H., Katoh, E., Luchi, S., Kobayashi, M., Maeda, T., Matsuoka, M., Yamaguchi, I. 2006. Identification and characterization of *Arabidopsis* gibberellin receptors. *Plant Journal*, 46, 880-889.

Nakashima, K., Fujita, Y., Katsura, K., Maruyama, K., Narusaka, Y., Seki, M., Shinozaki, K., Yamaguchi-Shinozaki, K. 2006. Transcriptional regulation of ABI3- and ABA-responsive genes including RD29B and RD29A in seeds, germinating embryos, and seedlings of Arabidopsis. 2006. *Plant Molecular Biology*, 60, 51–68.

Nath, P., Trivedi, P.K., Sane, V.A., Sane, A.P. 2006. Role of ethylene in fruit ripening, in: N.A. Khan (Ed.), *Ethylene Action in Plants*, Springer-Verlag, Berlin, 2006, pp. 151–184.

Ni, Di-an, Yu Xiao-hong, Wang Ling-jian, Xu Zhi-hong. 2002. Aberrant development of pollen in transgenic tobacco expressing bacterial iaaM gene driven by pollen- and tapetum-specific promoters. *Acta Experimental Sinica*, 35, 1-6.

Nikolic, R., Mitic, N., Miletic, R. Neskovic, M. 2006. Effects of cytokinins on *in vitro* seed germination and early seedling morphogenesis in *Lotus corniculatus L. Journal of Plant Growth Regulation,* 25, 187–194.

Nonogaki, H. 2008. Repression of transcription factors by microRNA during seed germination and postgerminaiton. Another level of molecular repression in seeds. *Plant Signaling and Behavior*, 1, 65-67.

Pandey, S., Nelson, D., Assmann, S. 2009. Two novel GPCR-type G proteins are abscisic acid receptors in *Arabidopsis*. *Cell*, 136, 136–148.

Park, S., Fung, P., Nishimura, N., Jensen, D., Fujii, H., Zhao, Y., Lumb, S., Santiago, J., Rodrigues, A., Chow, T., Alfred, S., Bonetta, D., Finkelstein, R., Provart, N., Desveaux, D., Rodriguez, P., McCourt, P., Zhu, J., Schroeder, J., Volkman, B., Cutler, S. Abscisic acid inhibits type 2C protein phosphatases via the PYR/PYL family of START proteins. *Science*, 324, 1068-1071.

Penfield, S., Josse, E-M., Kannangara, R., Gilday, A.D. Halliday, K.J., Graham, I.A. 2005. Cold and light control seed germination through the bHLH transcription factor SPATULA. *Current Biology*, 15, 1998-2006.

Pennazio, S., Roggero, P. 1991. Effects of exogenous salicylate on basal and stress-induced ethylene formation in soybean, *Biologia Plantarum,* 33, 58–65.

Petruzzelli, L., Coraggio, I., Leubner-Metzger, G., 2000. Ethylene promotes ethylene biosynthesis during pea seed germination by positive feedback regulation of 1-aminocyclo-propane-1-carboxylic acid oxidase, *Planta*, 211, 144–149.

Petruzzelli L, Sturaro M, Mainieri D, Leubner-Metzger G. 2003. Calcium requirement for ethylene-dependent responses involving 1-aminocyclopropane-1-carboxylic acid oxidase in radicle tissues of germinated pea seeds. *Plant, Cell and Environment,* 26, 661–671.

Piccoli, P., Masciarelli, O., Bottini, R. 1996. Metabolism of 17,17-[2H2]-gibberellins A4, A9 and A20 by *Azospirillum lipoferum* in chemically-defined culture medium. *Symbiosis*, 21, 263.

Piskurewicz, U., Jikumaru, Y., Kinoshita, N., Nambara, E., Kamiya, Y., Lopez-Molina, L. 2008. The gibberellic acid signaling repressor RGL2 inhibits Arabidopsis seed germination by stimulating abscisic acid synthesis and ABI5 activity. *Plant Cell*, 20, 2729–2745.

Seed Germination and the Secondary Metabolites, Plant Hormones   93

Popko, J., Hänsch, R., Mendel, R., Polle, A., Teichmann, T. 2010. The role of abscisic acid and auxin in the response of poplar to abiotic stress. *Plant Biology*, 12, 242 – 258.

Rao, S.S.R., Vardhini, B.V., Sujatha, E., Anuradha, S. 2002. Brassinosteroids-a new class of phytohormones. *Current Science*, 82, 1239–1245.

Rashotte, A.M., Brady, S.R., Reed, R.C., Ante, S.J., Muday, G.K. 2000. Basipetal auxin transport is required for gravitropism in root of Arabidopsis. *Plant Physiology*, 122, 481-90.

Rashotte, A.M., Carson, S.D., To, J.P., Kieber, J.J. 2003. Expression profiling of cytokinins action in *Arabidopsis*. *Plant Physiology,* 132, 1998–2011.

Ritchie, S., Gilroy, S. 1998. Gibberellins: Regulating germination and growth. *New Phytologist,* 140, 363-383.

Riefler, M., Novak, O., Strnad, M., Schmulling, T. 2006. Arabidopsis cytokinin receptor mutants reveal functions in shoot growth, leaf senescence, seed size, germination, root development, and cytokinin metabolism *The Plant Cell*, 18, 40–54.

Rinaldi, L.M.R. 2000. Germination of seeds of olive (*Olea europea* L.) and ethylene production: effects of harvesting time and thidiazuron treatment, *Journal of Horticultural Science and Biotechnology*, 75, 727–732.

Sanchez, M., Gurusinghe, S., Bradford, K.J., Vazquez-Ramos, J. Differential response of PCNA and Cdk-A proteins and associated kinase activities to benzyladenine and abscisic acid during maize seed germination. 2005. *Journal of Experimental Botany*, 56, 515–523.

Santner, A., Calderon-Villalobos, L., Estelle, M. 2009. Plant hormones are versatile chemical regulators of plant growth. *Nature Chemical Biology,* 5, 301-307.

Schwechheimer, C. 2008. Understanding gibberellic acid signaling—are we there yet? *Current Opinion in Plant Biology,* 2008 11, 9–15.

Seo, M., Nambara, E., Choi, G., Yamaguchi, S. 2009. Interaction of light and hormone signals in germinating seeds. *Plant Molecular Biology,* 69, 463–472.

Shen, Y.Y., Wang, X.F., Wu, F.Q., Du, S.Y., Cao, Z., Shang, Y., Wang, X.L., Peng, C.C., Yu, X.C., Zhu, S.Y., Fan, R.C., Xu, Y.H., Zhang, D.P. 2006. The Mg-chelatase H subunit is an abscisic acid receptor. *Nature,* 443, 823–826.

Song, L., Ding, W., Zhao, M., Sun, B., Zhang, L., 2006. Nitric oxide protects against oxidative stress under heat stress in the calluses from two ecotypes of reed. *Plant Science.* 171, 449–458.

Spaepen, S., Versées, W., Gocke, D., Pohl, M., Steyaert, J., Vanderleyden, J. 2007. Characterization of phenylpyruvate decarboxylase, involved in auxin production of *Azospirillum brasilense*. *Journal of Bacteriology,* 189, 7626–7633.

Spaepen, S., Dobbelaere, S., Croonenborghs, A., Vanderleyden, J. 2008. Effects of *Azospirillum brasilense* indole-3-acetic acid production on inoculated wheat plants. *Plant and Soil.* 312, 15–23.

Tian, X., Lei, Y., 2006. Nitric oxide treatment alleviates drought stress in wheat seedlings. *Biologia Plantarum,* 50, 775–778.

Tiedemann, J., Neubohn, B., Muntz, K. 2000. Different functions of vicilin and legumin are reflected in the histopattern of globulin mobilization during germination of vetch (*Vicia sativa* L.) *Planta,* 211, 1-12.

Tseng, M.J., Liu, C.W., Yiu, J.C., 2007. Enhanced tolerance to sulfur dioxide and salt stress of transgenic Chinese cabbage plants expressing both superoxide dismutase and catalase in chloroplasts. *Plant Physiology and Biochemistry,* 45, 822–833.

To, J.P., Kieber, J.J. 2008. Cytokinin signaling: two-components and more. *Trends in Plant Science,* 13, 85-92.

Toyomasu, T., Yamane, H., Murofushi, N., Inoue. Y. 1994. Effects of exogenously applied gibberellin and red light on the endogenous levels of abscisic acid in photoblastic lettuce seeds, *Plant Cell Physiology,* 35,127–129.

Ueguchi-Tanaka, M. Ashikari, M., Nakajima, M., Itoh, H., Katoh, E., Kobayashi, M., Chow, T.Y., Hsing, Y.I., Kitano, H., Yamaguchi, H., Matsuoka, M. 2005. Gibberellin Insensitive Dwarf1 encodes a soluble receptor for gibberellin. *Nature,* 437, 693-698.

Urao, T., Yamaguchi-Shinozaki, K. and Shinozaki, K. 2000. Two-component systems in plant signal transduction. *Trends in Plant Science,* 5, 67-75.

Vande Broek, A., Lambrecht, M., Eggermont, K., Vanderleyden, J. 1999. Auxins up-regulate expression of the indole-3-pyruvate decarboxylase gene from Azospirillum brasilense. *Journal of Bacteriology*, 181, 1338–1342.

Vande Broek, A., Gysegom, P., Ona, O., Hendrickx, N., Prinsen, E., Van Impe, J., Vanderleyden, J. 2005. Transcriptional analysis of the Azospirillum brasilense indole-3-pyruvate decarboxylase gene and identification of a cis-acting sequence involved in auxin responsive expression. *Molecular Plant–Microbe Interactions,* 18, 311–323.

Vandendussche, F., Van Der Straeten, D. 2007.Cross-talk of multiple signals controlling the plant phenotype, *Journal of Plant Growth Regulation*, 26, 176–187.

Versées, W., Spaepen, S., Vanderleyden, J., Steyaert, J. 2007a. The crystal structure of phenylpyruvate decarboxylase from *Azospirillum brasilense* at 1.5 Å resolution—implications for its catalytic and regulatory mechanism. *FEBS Journal*, 274, 2363–2375.

Versées, W., Spaepen, S., Wood, M.D., Leeper, F.J., Vanderleyden, J., Steyaert, J. 2007b. Molecular mechanism of allosteric substrate activation

Seed Germination and the Secondary Metabolites, Plant Hormones 95

in a thiamine diphosphate-dependent decarboxylase. *Journal of Biological Chemistry*, 282, 35269–35278.

Wang, Z.Y., Seto, H., Fujioka, S., Yoshida, S., Chory, J. 2001. BRI1 is a critical component of a plasma-membrane receptor for plant steroids. *Nature*, 410, 380–383.

Wang, X., Li, X., Meisenhelder, J., Hunter, T., Yoshida, S., Asami, T., Chory, J. 2005b. Autoregulation and homodimerization are involved in the activation of the plant steroid receptor BRI1. *Developmental Cell,* 8, 855–865.

Wang, X., Chory, J. 2006. Brassinosteroids regulate dissociation of BKI1, a negative regulator of BRI1 signaling, from the plasma membrane. *Science,* 313, 1118–1122.

Wang, A.X., Wang, X.F., Ren, Y.F., Gong, X.M., Bewley. J.D. 2005a. Endo-bmannanase and b-mannosidase activities in rice grains during and following germination, and the influence of gibberellin and abscisic acid, *Seed Science Research*, 15, 219–227.

Werner, T., Motyka, V., Strnad, M., Schmulling, T. 2001. Regulation of plant growth by cytokinin. Proceeding of the National Academic of Sciences USA 98, 10487–10492.

Weyers, J.D.B., Paterson, N.W. 2001. Plant hormones and the control of physiological processes New Phytologist 152, 375–407.

Willige, B., Ghosh, S., Nill, C., Zourelidou, M., Dohmann, E., Maier, A., Schwechheimer, C. 2007. The DELLA domain of GA Insensitive mediates the interaction with the GA Insensitive DWARF1A gibberellin receptor of *Arabidopsis. Plant Cell,* 19, 1209-1220.

White, C.N., Proebsting, W.M., Hedden, P., Rivin, C.J. 2000. Gibberellins and seed development in maize I. Evidence that gibberellin/abscisic acid balance governs germination versus maturation pathways, *Plant Physiology*, 122, 1081–1088.

White, C.N., Rivin, C.J. 2000. Gibberellins and seed development in maize. II. Gibberellin synthesis inhibition enhances abscisic acid signaling in cultured embryos, *Plant Physiology*, 122, 1089–1097.

Wilson, K.A. 1986. Role of proteolytic enzymes in the mobilization of protein reserves in germinating dicot seeds. In: Dalling, M.J., ed. *Plant Proteolic Enzymes*, Vol. II. Boca Raton, Florida: CRC Press Inc., 20-47.

Xu, Zhi-hong, Ni, Di-an. 1999. Modifications of leaf morphogenesis induced by inhibition of auxin polar transport. In: Altman A. et al.(eds.), *Plant Biotechnology and In Vitro Biology in the 21th Century*, Kluwer Academic Publ., Dordrecht, pp.97.

Yamada, H., Suzuki, T., Terada, K., Takei, K., Ishikawa, K., Miwa, K., and Mizuno, T. 2001. The Arabidopsis AHK4 histidine kinase is a cytokinin-binding receptor that transduces cytokinin signals across the membrane. *Plant Cell Physiology*, 42, 1017–1023.

Yamaguchi, S. 2008. Gibberellin metabolism and its regulation. *Annual Review of Plant Biology*, 59, 225-251.

Yang, S.F., Hoffman, N.E. 1984. Ethylene biosynthesis and its regulation in higher plants. *Annual Reviews of Plant Physiology,* 35, 155–189.

Yin, Y., Wang, Z., Mora-Garcia, S., Li, J., Yoshida, S., Asami, T., Chory, J. 2002. BES1 accumulates in the nucleus in response to brassinosteroids to regulate gene expression and promote stem elongation. *Cell,* 109, 181–191.

Zapata, P.J., Serrano, M., Pretel, M.T., Amorós, A., Botella, M.A. 2004. Polyamines and ethylene changes during germination of different plant species under salinity. *Plant Science,* 167, 781–788.

Zhang, H., Shen,W.B., Zhang,W.,X u, L.L. 2005. A rapid response of β-amylase to nitric oxide but not gibberellin in wheat seeds during the early stage of germination. *Planta,* 220, 708–716.

Zhang, S., Cai, Z., Wang, X. 2009. The primary signaling outputs of brassinosteroids are regulated by abscisic acid signaling. Proceedings of the National Academy of Sciences, USA 1106:4543-4548.

Zheng, C., Jiang, D., Liub, F., Dai, T., Liu, W., Jing, Q., Cao, W. 2009. Exogenous nitric oxide improves seed germination in wheat against mitochondrial oxidative damage induced by high salinity. *Environmental and Experimental Botany*, 67, 222-227.

Zimmer, W., Bothe, H. 1988. The phytohormonal interactions between *Azospirillum* and wheat. *Plant and Soil,* 110, 239–247.

In: Flowering Plants
Editor: Jeremy J. Tellstone

ISBN: 978-1-61324-653-5
© 2011 Nova Science Publishers, Inc.

Chapter 4

RISE OF THE CLONES: APOMIXIS IN PLANT BREEDING

*Pablo Bolaños-Villegas[1,2,3], Saminathan Thangasamy[1,2,3] and Guang-Yuh Jauh[*1,3,4]*

[1]Taiwan International Graduate Program,
Molecular and Biological Agricultural Sciences, Academia Sinica,
Taipei 11529, Taiwan, R. O. C.
[2]Graduate Institute of Biotechnology, National Chung-Hsing University,
Taichung 402, Taiwan, R.O.C.
[3]Institute of Plant and Microbial Biology, Academia Sinica,
Taipei 11529, Taiwan, R.O.C.
[4]Biotechnology Center, National Chung-Hsing University,
Taichung 402, Taiwan, R. O. C.

ABSTRACT

In flowering plants, the transfer of traits from one generation to another is accomplished by fertilization of the female gametophyte with sperm cells delivered by the pollen tube and subsequent reassortment of traits (alleles) in the developing progeny. In nature, DNA recombination

---

[*] Corresponding author: Guang-Yuh Jauh Institute of Plant and Microbial Biology, Academia Sinica Nankang, Taipei, Taiwan 11529, Tel: 886-2-27871156, Fax: 886-2-27827954, E-mail: jauh@gate.sinica.edu.tw.

and segregation of traits in the progeny prevent the accumulation of deleterious genes and loss of fitness; however, for breeding purposes, fixing superior trait combinations (genotypes) and circumventing sexual reproduction is advantageous. The formation of sexual seed without fertilization of the egg is called apomixis and is considered the "holy grail" of plant breeding because top-performing varieties can be reproduced indefinitely without changes in the genotypes themselves or their expression patterns. Apomixis is a dominant trait and consists of several processes working in tandem; separately, each process is detrimental for plant reproduction, but as a single unit, they allow development of embryos and endosperm from unfertilized eggs. The 3 processes of apomixis include 1) apomeiosis, or cell division without DNA recombination in pollen and eggs; 2) parthenogenesis, autonomous development of eggs into fully formed embryos; and 3) stable development of the endosperm, the part of the seed needed for the embryo to grow. Unfortunately, full expression and transmission of apomixis is affected by DNA recombination; therefore, introgression of the trait from wild relatives into commercial varieties is extremely difficult. In this review, we report on genes that have been identified to regulate each step of apomixis and discuss strategies to allow full transmission of the trait with tools from molecular biology.

CHALLENGES FOR MODERN AGRICULTURE

The world's population is believed to grow to 9.1 billion by 2050 (Editorial, 2010b), roughly a 30% increase in the number of people that need to be fed. With the agricultural tools available today, about 1 billion people are still undernourished, mostly those in Sub-Saharan Africa (265 million people) and Southeast Asia (642 million) (Editorial, 2010b; Williams, 2010). An increase in agricultural output is necessary and could be attained in two ways: 1) increase the area of cultivated land, tax endangered ecosystems and decimate any existing biodiversity; or 2) intensify productivity in a sustainable manner (Editorial, 2010a). In both scenarios, crops may still have to deal with global environmental change caused by an increase in CO_2 emissions (Woodward et al., 2009). Therefore, a deep understanding of the latest developments in plant biology research is required to devise ways to increase food production.

Several alternatives proposed to boost plant productivity include improved nitrogen use efficiency, increased water use efficiency, introduction of new genes to enhance defense against pests, and apomixis (Pennisi, 2010).

Apomixis is an asexual form of seed formation that does not affect seed viability (Hare, 2008; Tucker and Koltunow, 2009). Apomixis is advantageous because it fixates hybrid vigor (heterosis), which, in crops such as maize, leads to increasing yield by up to 20% (Siddiqi et al., 2009). In most crops, hybrid vigor is not stable and declines steadily after each new generation because of the segregation of beneficial alleles at the time of sexual reproduction. Therefore, and at least in theory, apomixis would allow for the fixation and propagation of any elite genotype (Hwa and Yang, 2008; Siddiqi et al., 2009) while avoiding complications such as self-incompatibility or the transfer of viruses, because it occurs in vegetatively propagated plants (Bicknell and Koltunow, 2004). In developing countries, the availability of apomictic seed would reduce the dependence of farmers on commercial seed because inbreeding effects (the reduction of heterosis) would no longer occur, and the seed could be re-used season after season.

BIOLOGICAL BASIS OF APOMIXIS

Apomixis is in itself a modified version of sexual reproduction in which the maternal embryo originates from a diploid (2n) egg cell without the contribution of a male gamete (Grimanelli et al., 2001). During apomixis, sexual recombination (e.g., meiosis) is omitted, which results in female gametophytes and gametes with a genetic composition and chromosome number identical to the mother. Then these gametes initiate embryogenesis and endosperm proliferation without fertilization by foreign pollen (Albertini et al., 2010). Apomixis has been reported to occur naturally in *Tripsacum dactyloides*, a wild relative of maize (*Zea mays*); *Pennisetum squamulatum*, a relative of pearl millet (*P. glaucum*); *Sorghum bicolor*, an important millet in tropical regions (Carman et al., 2011); and several other genera such as *Panicum* (Yamada-Akiyama et al., 2009), *Paspalum, Taraxacum, Eulaliopsis* (Yao et al., 2007), *Pilosella aurantiaca* (Krahulcova et al., 2010) and *Hieracium*. However, the trait is absent in major crops (Spillane et al., 2004; Catanach et al., 2006). The trait is dominant and is inherited in a Mendelian fashion, but efforts to transfer it through hybridization into agriculturally important species have failed (Albertini et al., 2010). One reason for the failure is that apomixis is the result of 3 separate processes -- meiosis, parthenogenesis and functional endosperm formation -- which must "short-circuit" in tandem simultaneously (Grimanelli et al., 2001)(Figure 1). Apomixis may result from global regulatory changes that include the sexual developmental pathway, and these changes may occur as a consequence of hybridization and polyploidization, because most apomicts in nature are polyploid (Sharbel et al., 2009; Baubec et al., 2010). Apomictic traits are unlikely to evolve randomly because mutants for each process would show reduced fitness and would be eliminated rapidly from the population (van Dijk, 2009). Above all, for apomixis to have an impact on yield, obligate instead of facultative apomixis must be identified (Sorensen et al., 2009). Once all key genes have been identified and isolated, new breeding technologies will be adapted and developed to capitalize on this knowledge because conventional approaches have not been successful in creating apomictic crops.

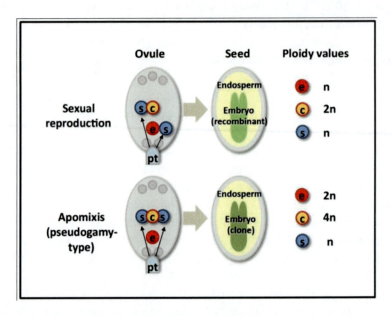

Figure 1. Ovule and seed development in sexual and apomictic plants. In plants that reproduce sexually, the ovule may contain 8 nuclei (Polygonum-type ovule), including the egg (e) and the central cell (c). The egg is haploid (n), and the central cell is diploid (2n). Both are targeted by sperm cells(s) delivered by a pollen tube (pt). The result is the fusion of their DNA. The fertilized egg develops into a diploid (2n) embryo, and the fertilized central cell develops into triploid (3n) endosperm. The egg and the sperm cells are the product of DNA recombination during meiosis. Thus, sexual embryos are not clones of either parent; they are recombinant. In contrast, in apomictic plants, the egg mostly bypasses meiosis (apomeiosis) and is diploid. Diploid eggs may initiate autonomous development into embryos (parthenogenesis). Only the central cell (2n to 4n) is fertilized by 2 sperm cells (pseudogamy), although it may proliferate autonomously as well. The result is an embryo (2n) that fully resembles the maternal plant and is surrounded by functional endosperm (6n).

In *Pennisetum* and *Paspalum*, fluorescent *in situ* hybridization (FISH) has revealed that loci for apomeiosis and parthenogenesis cluster in large chromosomal regions (50–129 Mbp) named apospory-specific genomic regions (ASGRs) or apomixis-controlling loci (ACLs) (Akiyama et al., 2004; Calderini et al., 2006; Ozias-Akins and van Dijk, 2007). These regions are rich in heterochromatin (transcriptionally inactive chromatin) and are resistant to recombination (Catanach et al., 2006). The lack of recombination is attributed to chromosomal rearrangements that lead to the evolution of functionally haploid alleles (hemizygous alleles). The spread of heterochromatin into ASGRs may preserve linkage in these groups of alleles while preventing the activation of harmful transposable elements (Martienssen, 2010). In practice, the existence of heterochromatic ASGRs means that apomictic loci do not hybridize well to restriction fragment length polymorphism (RFLP) or amplified length polymorphism (AFLP) markers, so estimating precisely the location of apomictic loci for subsequent cloning is difficult (Ozias-Akins and van Dijk, 2007). As well, because of the lack of recombination in apomictic regions, fully introgressing the trait from apomictic relatives into commercial crops is difficult (Leblanc et al., 2009). Other limitations include differences in ploidy and self-incompatibility (Horandl and Temsch, 2009).

An important component of apomixis is modified endosperm development, which has been studied extensively (Vielle-Calzada et al., 2000; Berger and Chaudhury, 2009).

In most flowering plants, endosperm development is triggered by the fusion of a haploid (1n) sperm cell with a diploid (2n) central cell within the ovule sac during double fertilization. This fusion translates into a parental transmission ratio of 2m (m for maternal) to 1p (p for paternal) (Köhler and Grossniklaus, 2005). Deviation from this ratio because of differences in ploidy or defects in fertilization usually leads to seed underdevelopment (with the paternal ratio < 1) (Erilova et al., 2009). Apomictic plants are believed to show increased tolerance to deviation from the 2m + p ratio because of changes in the expression of genes that mediate the silencing of paternal genes in the endosperm (Albertini et al., 2010).

The silencing of paternal genes is known as imprinting; it is mostly the result of differences in DNA/histone methylation and may regulate competition between parental enhancers and maternal inhibitors of embryo growth (PEGs and MIGs) (Jullien et al., 2008; Jullien and Berger, 2010). Interestingly, most apomictic plants require fertilization of the central cell (pseudogamy) (Ozias-Akins and van Dijk, 2007), perhaps as a mechanism to comply with the 2m + p ratio of parental gene dosage even if the progeny is clonal.

Several modifications to the general plan of apomixis exist and need to be understood. The most common include the following.

a) Apospory: a diploid embryo sac develops from a somatic nucellar cell, which then develops as a functional megaspore, sometimes called an aposporous initial cell (the nucellus is the area within the ovule that gives rise to germ cells) (Ozias-Akins and van Dijk, 2007; Tucker and Koltunow, 2009). Both the aposporous initial cell and the original megaspore may be formed at the same time within the ovule, thus leading to intra-ovular competition and sometimes polyembryony (Tucker and Koltunow, 2009).

b) Mitotic diplospory: the embryo sac develops from a megaspore mother cell that underwent modified meiosis. If the suppression of meiosis is complete, the fixation of the maternal genome is guaranteed (mitotic-diplospory) (Tucker and Koltunow, 2009; Albertini et al., 2010).

c) Aneuspory: the fixation of the maternal genome is not guaranteed if meiosis begins but homologous chromosomes show reduced pairing. As a result, recombination of maternal DNA may occur, albeit at a low frequency (Albertini et al., 2010).

d) Autonomous endospermy: the endosperm is initiated autonomously without contribution from sperm cells. The ploidy of the endosperm may vary depending on the frequency at which haploid polar nuclei fuse with each other before the endosperm starts to form. Male organs may not be functional in this type of apomixis (Albertini et al., 2010).

e) Pseudogamy: in this type of apomixis, the diploid central cell (the result of the fusion of polar nuclei) must be fertilized by a sperm cell to generate viable and functional endosperm (Spillane et al., 2004).

TECHNIQUES TO STUDY APOMICTIC GENES

The characterization of apomixis at the molecular level has involved two different approaches. The first approach is to use plants that reproduce sexually, such as Arabidopsis thaliana. These plants are used as model organisms in which mutations are induced by chemicals (ethyl methyl sulfonate; EMS), physical agents (γ-rays and fast neutrons), or exposure to

Agrobacterium tumefaciens, a bacterium that inserts T-DNA segments randomly into the target genome (Page and Grossniklaus, 2002). The use of EMS induces gain-of-function mutations, which are useful because apomixis as a whole behaves as a dominant trait (Albertini et al., 2010). However, exposure to A. tumefaciens creates loss of function, and then, genes can be studied individually by comparing the phenotype caused by recessive alleles with that observed in the wild type (Page and Grossniklaus, 2002). Both approaches have been useful in dissecting the function of genes that may mediate apomeiosis, parthenogenesis and modified endosperm development (Albertini et al., 2010) (Table 1). Some of these genes induce somatic embryogenesis, such as Somatic Embryogenesis Receptor-Like Kinase 1 (SERK 1), which codes for a receptor-like kinase (Hecht et al., 2001); Wuschel (WUS), which codes for a homeodomain protein involved in the establishment of polarity in embryos (Su et al., 2009; Smith and Long, 2010); and Leafy Cotyledon 1/2 (LEC1/2), a transcription factor (Stone et al., 2008). Research into apomictic crops, such as guineagrass, has also created a wealth of expressed sequence tags (ESTs), isolated from pistils (Yamada-Akiyama et al., 2009), which will provide useful information in the engineering of apomixis.

Other genes mediating key aspects of DNA recombination and cell cycle regulation during meiosis include Spo homolog 1/2 (SPO11-1/2), a topoisomerase; Recombination8 (REC8) and DYAD/Switch1 (DYAD), two regulators of chromosome cohesion; and Omission of Second Division (OSD), a regulator of entry into meiosis II. When SPO11-1, REC8 and OSD1 are knocked out simultaneously, meiosis no longer involves recombination and resembles mitosis instead, although this leads to continuous polyploidization (d'Erfurth et al., 2010). This genotype is called the mitosis instead of meiosis (MiMe) genotype and has potential applications in crop improvement (d'Erfurth et al., 2009). Mutations in DYAD are also interesting; female gamete formation occurs without the need for meiosis (Ravi et al., 2008; Chan, 2010).

Study of Arabidopsis has also been extremely useful in investigating imprinting and endosperm development; for instance, it has been possible to determine that the polycomb group complex (PcG) mediates the differential silencing of genes that promote endosperm development (Berger and Chaudhury, 2009). Some of these genes are Fertilization Independent Seed 2 (FIS2), a gene that codes for a zinc finger protein and mediates seed growth (Jullien et al., 2006); Fertilization Independent Endosperm (FIE), which codes for a WD-40 domain protein involved in endosperm proliferation (Yadegari et al., 2000); and MEDEA (MEA), a gene for SET-domain protein that behaves as a ploidy sensor during gene dosage regulation (Erilova et al., 2009). These genes interact with several modulators, including DNA Methyltransferase 1 (MET1), which represses the expression of the paternal FIS2 locus (Luo et al., 1999); Demeter (DME), a demethylase that activates transcription of MEA (Choi et al., 2002; Tiwari et al., 2008); and Multicopy Supressor of IRA 1 (MSI1), which silences the expression of MET1 itself (Jullien et al., 2008). Methyltransferases are also key regulators of apospory and apomeiosis in maize, as shown by maize–Tripsacum hybrids that show loss of expression of the genes DMT102/103 (Garcia-Aguilar et al., 2010). In fact, deep gene expression analysis of the apomict diploid Boechera holboelli indicated that genes for the PcG are upregulated in late stages of ovule development (Sharbel et al., 2010). In recent years, apomixis-related genes have been identified through proteomics. Such an approach currently involves the comparison of 2D protein profiles from different reproductive stages in apomictic plants, then sequencing specific proteins (Li et al., 2009). Proteomics could accelerate the identification of apomictic genes in angiosperms (Miernyk et al., 2010).

Investigation of apomictic plants has involved use of molecular markers to locate and clone loci that regulate apomixis, although their use is hampered by the resistance to recombination found in ASGRs (Pupilli et al., 2001). To circumvent this problem, study of the apomictic genus *Hieracium* has involved irradiation with radioactive cobalt (^{60}Co) at 40 krad, a median lethal dose (LD50) for dry seed (Catanach et al., 2006). Previously sequenced AFLP markers were investigated in mutants that showed discrete loss of apomeiosis and parthenogenesis.

The analysis is straightforward and intuitive: loss of AFLP markers plus loss of apomeiosis or parthenogenesis would indicate that a particular locus has a role in apomixis and that the lost markers may reside adjacent to apomictic loci (Catanach et al., 2006). Use of microsatellite markers has been beneficial for identifying apomictic traits in sibling populations of cassava (Nassar et al., 2009). Another approach to study apomixis involves flow cytometry of seeds collected from plants in which apomixis is suspected to occur (Matz et al., 2000). The rationale is that in sexual individuals, seeds composed of a 2C embryo and 3C endosperm develop, which reflects a parental genome composition of 1C paternal and 1C maternal for the embryo and 2C maternal plus 1C paternal for the endosperm (Sharbel et al., 2009).

Table 1. Genes involved in apomixes

Gene name	Organism	Function	Process regulated	Source
LEAFY COTYLEDON 2 (LEC2)	*Arabidopsis*	B3-domain transcription factor	Embryogenesis	Stone *et al.*, 2008
SERK1	*Arabidopsis*	Receptor-like kinase	Embryogenesis	Hecht *et al.*, 2007
WUSCHEL	*Arabidopsis*	Homeodomain protein, shoot apical meristem initiation	Embryogenesis	Smith *et al.*, 2011; Su *et al.*, 2009
OSD	*Arabidopsis*	Regulation of meiotic progression	Male and female apomeiosis	D'Erfurth *et al.*, 2010
REC8	*Arabidopsis*	Regulation of sister chromatid cohesion	Male and female apomeiosis	D'Erfurth *et al.*, 2010
SPO11-1	*Arabidopsis*	Formation of double strand breaks in DNA	Male and female apomeiosis	D'Erfurth *et al.*, 2010
DYAD/ SWITCH1 (SWI1)	*Arabidopsis*	Regulation of sister chromatid cohesion	Female apomeiosis	Ravi *et al.*, 2008 (no reference)
FERTILIZATION INDEPENDENT ENDOSPERM (FIE)	*Arabidopsis*	WD40-domain protein, polycomb complex protein, DNA methylation	Autonomous endosperm proliferation	Yadegari *et al.*, 2000
FERTILIZATION INDEPENDENT SEED 2 (FIS2)	*Arabidopsis*	C2H2-zinc finger protein, polycomb complex protein	Autonomous endosperm proliferation	Jullien *et al.*, 2006
MEDEA (MEA)	*Arabidopsis*	SET-domain polycomb complex protein, DNA methylation	Parental gene dosage in endosperm	Erilova *et al.*, 2009
MET1	*Arabidopsis*	DNA methyltransferase, silencing of *FIS2*	Parental gene dosage in endosperm	Luo *et al.*, 1999
MULTICOPY SUPRESSOR OF IRA 1 (MSI1)	*Arabidopsis*	Polycomb complex protein, repression of expression of *MET1*	Autonomous endosperm proliferation	Jullien *et al.*, 2008
DEMETER (DME)	*Arabidopsis*	DNA glycosylase, required for the expression of *MEA*		Choi *et al.*, 2002; Tiwari *et al.*, 2008
DNA METHYL-TRANSFERASE 102 (DMT102)	Maize	Maintenance of DNA methylation	Parthenogenesis	García-Aguilar *et al.*, 2010
DNA METHYL-TRANSFERASE 103(DMT103)	Maize	*De novo* DNA methylation	Parthenogenesis, apomeiosis	García-Aguilar *et al.*, 2010

In contrast, apomictic diploids mostly show embryos with a 2C composition, probably because of apomeiosis, and endosperm with a composition of 6C, possibly 4C of maternal origin (from the central cell) and 2C paternal, probably from 2 sperm cells (Sharbel et al., 2009). This approach consists of the use of mature seed, without any pretreatment. Seeds are chopped and stained with 4,6-diamidino-2-phenylindole (DAPI) for direct analysis in a flow cytometer (Matz et al., 2000). Results are not influenced by inclusion of the seed coat because the seed coat contains no nuclei. Also, with the approach, differentiating the endosperm from the embryo is facilitated because the endosperm contains less nuclei, which is reflected by sharp differences in the height and position of the peaks on flow cytometry (Matz et al., 2000).

Once apomixis is found to occur in a particular material, changes in gene expression patterns could be compared throughout several flower development stages by means of serial analysis of gene expression (SAGE), a method that quantitates the abundance of mRNA molecules (Sharbel et al., 2010). For instance, results from SAGE studies of *Boechera holboelli* suggest that shifts in the time of expression (heterochrony) of both imprinted and non-imprinted genes may bring about apomixis (Sharbel et al., 2010), which FISH results suggest may be regulated by the largely heterochromatic *Het* chromosome (Kantama et al., 2007).

TECHNICAL BARRIERS TO THE STUDY OF APOMIXIS

Characterization of apomictic genes and loci in relevant crops is hoped to lead to the full engineering of apomixis; however, as discussed before, apomixis depends on a complex array of alleles that must be deregulated at the same time (van Dijk, 2009). Only a few genes are believed to be required to induce apomixis; however, this switch needs to be sustained by profound changes in the reproductive biology of a plant, including deep epigenetic reprogramming (Leblanc et al., 2009). In essence, whether single, double or triple mutant backgrounds can be created through re-transformation or continuous crossing and second mutant alleles inserted randomly is unclear, and since the alleles are not linked, they will segregate (Halpin, 2005). Therefore, gene stacking may contribute to apomictic-like phenotypes in only sexual crops, which limits the usefulness of this approach in molecular plant breeding (Figure 2). Emerging technologies that can speed up the engineering of transgene cassettes and fully trigger each of the three components of apomixis must be considered important (Spillane et al., 2004).

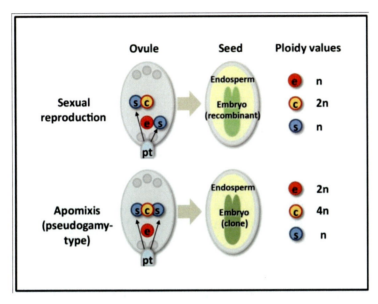

Figure 2. 1-4. Proposed methods to engineer apomixis in plants. 1) Introgression of apospory genomic regions (ASGRs), also named apomixis-controlling loci (ACL) from wild relatives into crop species requires that ASGRs are inherited to mixed progenies owing to recombination during meiosis. ASGRs are large complexes of linked genes that regulate apomeiosis, parthenogenesis and modified endosperm development. ASGRs are resistant to recombination; therefore, the progenies seldom inherit apomictic-like traits. 2) Gene stacking is the standard introduction of single transgenes into plants to induce a trait. It may be used to deliver a) T-DNA to turn off a gene or b) a transgene to induce a novel function. Gene stacking has been used in the model plant Arabidopsis to induce traits such as apomeiosis. Insertion of the genes is random and leads to segregation in the progenies. Repeated cycles of transformation or cross-pollination are required to accumulate transgene complexes. 3) Minichromosome engineering refers to the truncation of a chromosome arm, which also receives a specific Cre/lox recombination site for the controlled addition of transgenes. In theory, large amounts of DNA cloned into appropriate vectors could be added sequentially to create complexes of linked genes. This approach is available only in maize. 4) Targeted genome elimination replaces an important protein found in the centromere, thus causing controlled missegregation and loss of chromosomes from either parent. In crosses, one of the parents is transgenic and the other is the wild type for the centromeric protein. During the cross, all the transgenic chromosomes are eliminated. Because the wild type undergoes meiosis, the progeny is recombinant. This approach is purely experimental and has been tested only in Arabidopsis. Blue is wild type; orange is apomictic locus or transgene.

New Approaches to the Study of Apomixia

A. Minichromosome Engineering

The introduction of several genes at the same time has attracted great attention. Simultaneous multigene engineering in plants may be required to engineer traits that depend on the expression of quantitative trait loci (QTL) or a stable multiple array of genes (SMART) locus, which contains multiple genes and is stable through meiosis (Naqvi et al., 2009).

We can stretch the definition of an ASGR as a SMART locus and consider different techniques to deliver a SMART locus of 50 Mbps. *Agrobacterium* high-capacity vectors such as binary bacterial artificial chromosomes (BIBACs) or transformation-competent artificial chromosomes (TACs) have capacity in the range of 200 kbp, the same range attained with the use of bioactive beads coated with DNA (Naqvi et al., 2009). BIBACs and TACs may not deliver an entire ASGR; however, the study of maize has contributed to the development of plant artificial minichromosomes based on maize B chromosomes (Pennisi, 2010). This technology has created the hope that in the near future, stable environments may be created for the targeted addition of unlimited amounts of DNA together with their regulatory elements, thus allowing persistent expression of gene complexes (Houben and Schubert, 2007; Yu et al., 2007). Maize B chromosomes are extra chromosomes that are mostly transcriptionally inert and possess an accumulation mechanism that allows them to segregate exclusively into sperm cells after pollen mitosis II (Jones et al., 2008). Moreover, such chromosomes contain a chromosome-specific DNA repeat around the centromere, which facilitates their identification (Birchler et al., 2010). Minichromosomes are created by truncating B chromosomes, accomplished by transforming plants with vectors carrying a) a herbicide resistance gene, b) a Cre/lox recombination site, c) a sequence encoding a green fluorescent protein (GFP) or a red fluorescent protein and, d) telomere repeats (Yu et al., 2007). After transfection or bombardment into early maize embryos, the transformed DNA is inserted through nonhomologous end joining into chromosomal areas where DNA has been broken. This insertion is possible because the telomere sequence in the vector is enough to attract the machinery involved in nonhomologous end joining (Birchler et al., 2010). Insertion of the vector DNA occurs mostly in the distal portion of a chromosome, effectively cleaving off a chromosome arm and substituting the original arm fragment with a telomere (Birchler et al., 2010). Insertion is random, and *in situ* hybridization with a chromosome-specific marker is required to confirm the identity of the truncated chromosome (Houben and Schubert, 2007). This entire process is named "telomere capping" and was first developed by Farr and associates (Farr et al., 1992) to create human artificial chromosomes with functional centromeres. Centromeres are loci that interact with kinetochores, protein complexes that bind to spindle microtubules and regulate chromosome segregation during cell division (Chan, 2010).

In plants, functional centromeres are not possible to recreate *de novo* because transformation with a primary DNA sequence alone leads to the loss of the centromeric histone protein CENH3, which eventually leads to inactivation of the centromere (Phan et al., 2007; Han et al., 2009).

For minichromosome engineering, inert B chromosomes are preferred targets for truncation because their survival rate is higher (Birchler et al., 2010). Truncation of normal A chromosomes is also possible but requires the embryo to have a polyploid (tetraploid) or aneuploid background (Houben and Schubert, 2007). Plants that carry the truncated chromosome should be recognized cytologically first, including an assay for the presence of CENH3 in the truncated chromosome, and then crossed to a diploid plant for 2 generations to obtain a diploid addition line (Houben and Schubert, 2007; Birchler et al., 2010).

Once the addition line is created, it could be transformed with an appropriate vector carrying the target genomic fragment, in this case a bacterial artificial chromosomes carrying a portion of a previously cloned ASGR plus a Cre/lox recombination site (Birchler et al., 2010) (Figure 2). This approach appears promising, but several issues remain unsolved. For instance, the truncation efficiency is low (Houben and Schubert, 2007), and more efficient site-specific integration systems might be required, such as the use of zinc-finger nucleases, which are engineered to recognize a target sequence, induce double-strand breaks in the sequence and allow insertion of donor plasmids by homologous recombination (Shukla et al., 2009; Birchler et al., 2010).

Also, because ASGRs are heavily heterochromatic, whether their addition into minichromosomes induces loss of the original epigenetic state is unclear. One possibility is to create chimeric constructs containing key genes (Halpin, 2005), in this case of apomixis, and deliver them towards a minichromosome. To summarize, the uncertainties surrounding minichromosome technology to engineer apomixis suggest that this method should be approached with great caution and skepticism.

B. Targeted Genome Elimination

Targeted genome elimination is reported to occur naturally in crosses between cultivated barley (*Hordeum vulgare*) and *H. bulbosum*. In the resulting embryos, chromosomes inherited from *H. bulbosum* missegregate and are lost (Kasha and Kao, 1970). The mechanism that regulates this behavior is unknown (Barret et al., 2008); however, work in *Arabidopsis* suggests that subtle modifications in the sequence that codes for the centromere histone CENH3 may induce missegregation and loss of chromosomes in plants harboring the mutation (Ravi and Chan, 2010). Loss of modified chromosomes occurs only during mitosis within the zygote and leads to inheritance of wild-type chromosomes, thus leading to progenies with recombinant versions of either parent, independent of the genome composition (Figure 2).

The ploidy of the progenies varies, with approximately 50% being haploid and the rest either diploid or aneuploid because of spontaneous non-reduction of gametes (Ravi and Chan, 2010). The rationale behind this method is complex. As mentioned earlier, centromeres are the chromosomal loci that attach to spindle microtubules to mediate inheritance of the genome during cell division (Chan, 2010). These loci are epigenetically modified by incorporation of the histone HTR12, a variant of conventional H3 that is found in centromeric nucleosomes (Ravi and Chan, 2010). Histone HTR12 sequences evolve rapidly in their amino-terminal tail (N terminus); thus any change in the tail alters their specificity (Chan, 2010). In their original paper, Ravi and Chan (2010) described use of a homozygous *htr12* mutant in which they replaced the original N terminus with a sequence coding for a GFP-tag followed by another sequence encoding a conventional H3 sequence. The C-terminal histone fold domain remained unchanged. This construct is referred to as a GFP-tailswap and was more effective than replacement of the original HTR12 C terminus by conventional H3 (Ravi and Chan, 2010).

Although the approach does not induc apomixis *per se*, it has several advantages over other methods developed to maintain heterosis. First, it is applicable in all plant species (Ravi and Chan, 2010), regardless of the existence of apomixis in closely related species. Second, it eliminates any transgenes in the harvestable product, which is acceptable for consumers concerned with genetically modified organisms, a limiting factor for plant biotechnology in areas such as Europe (Spillane et al., 2004). Third, sexuality remains the default state, thus allowing incorporation of new genes through conventional plant breeding (Spillane et al., 2004). Fourth, use of GFP-tailswap plants as receptors eliminates transmission of transgenes into the wild. Finally, double haploid lines can be created with this approach, thus eliminating the need for tissue culture (Chan, 2010).

Improvements in the technique have been suggested, such as use of RNA interference (RNAi) in gametes to eliminate the expression of the native *HTR12* gene in pollen and egg cells. The same vector could express the GFP-tailswap construct with altered codon usage to escape RNAi. Therefore, targeted genome elimination lines could be engineered in one single transformation event (Chan, 2010). However, this experimental approach has been tested only in *Arabidopsis*, a plant with only 5 chromosomes, and whether the same results could be achieved in crops with more chromosomes is unclear (Copenhaver and Preuss, 2010).

C. Balancer Chromosome Technology

Conservation of hybrid vigor in plants may depend on counter-selection against homozygotes alone, without the development of apomixis (Chan, 2010). This approach is merely theoretical and is derived from the use of balancer chromosomes in mice. In mice, balancer chromosomes have an inversion that contains a recessive lethal mutation (Chan, 2010). Recombination leads to the formation of dicentric and acentric chromosomes that are fatal to gametes; therefore, this technology could be used to maintain specific alleles in heterozygous conditions. New translocations and inversions could be created with the help of site-specific recombinases such as those of the Cre/lox system (Gilbertson, 2003). This approach has a potential drawback in that it leads to reduced fertility. Therefore, it may be applied only to crops without seeds as a product, such as cassava. Nonetheless, because tissue culture is no longer required, related problems such as somaclonal variation are eliminated as well (Chan, 2010; Nassar et al., 2010).

PROSPECTS AND CONCLUSIONS

Apomixis holds the key to preserving complex heterozygous genotypes that sustainably increase agricultural output. In the face of continued population growth and the increasing need for food, this beneficial trait needs to be properly dissected to understand its regulatory mechanisms. Emerging technologies such as minichromosome engineering may one day allow for transmission of gene complexes in a stable chromosomal environment. However, more research is needed to translate these technologies into approaches that molecular plant breeders can use. Stronger collaboration and a better understanding between plant biologists, plant breeders and the private sector may lead to the engineering of apomixis into commercial crops in the near future. Therefore, apomixis has a key potential to revolutionize plant breeding and help implement sustainable agriculture.

ACKNOWLEDGMENTS

We appreciate Ms. Laura Smales for her excellent editing work. This work was supported by research grants from Academia Sinica (Taiwan), the National Science and Technology Program for Agricultural Biotechnology (NSTP/AB, 098S0030055-AA, Taiwan), the National Science Council (98-2313-B-001-001-MY3, 99-2321-B-001-036-MY3, Taiwan), and the Li Foundation (USA) to G.-Y. Jauh.

REFERENCES

Akiyama, Y., Conner, J.A., Goel, S., Morishige, D.T., Mullet, J.E., Hanna, W.W., and Ozias-Akins, P. (2004). High-resolution physical mapping in *Pennisetum squamulatum* reveals extensive chromosomal heteromorphism of the genomic region associated with apomixis. *Plant Physiology 134*, 1733-1741.

Albertini, E., Barcaccia, G., Mazzucato, A., Sharbel, T.F., and Falcinelli, M. (2010). Apomixis in the era of biotechnology. In: *Plant developmental biology - biotechnological perspectives*, E.C. Pua and M.R. Davey, eds (Berlin Heidelberg: Springer-Verlag), pp. 405-436.

Barret, P., Brinkmann, M., and Beckert, M. (2008). A major locus expressed in the male gametophyte with incomplete penetrance is responsible for *in situ* gynogenesis in maize. *Theoretical and Applied Genetics* 117, 581-594.

Baubec, T., Dinh, H.Q., Pecinka, A., Rakic, B., Rozhon, W., Wohlrab, B., von Haeseler, A., and Mittelsten Scheid, O. (2010). Cooperation of multiple chromatin modifications can generate unanticipated stability of epigenetic states in *Arabidopsis*. *The Plant Cell* 22, 34-47.

Berger, F., and Chaudhury, A.M. (2009). Parental memories shape seeds. *Trends in Plant Science* 14, 550-556.

Bicknell, R.A., and Koltunow, A.M. (2004). Understanding apomixis: recent advances and remaining conundrums. *Plant Cell 16 Suppl,* S228-245.

Birchler, J.A., Krishnaswamy, L., Gaeta, R.T., Masonbrink, R.E., and Zhao, C. (2010). Engineered minichromosomes in plants. *Critical Reviews in Plant Sciences* 29, 135-147.

Calderini, O., Chang, S.B., de Jong, H., Busti, A., Paolocci, F., Arcioni, S., de Vries, S.C., Abma-Henkens, M.H.C., Lankhorst, R.M., Donnison, I.S., and Pupilli, F. (2006). Molecular cytogenetics and DNA sequence analysis of an apomixis-linked BAC in *Paspalum simplex* reveal a non pericentromere location and partial microcolinearity with rice. *Theoretical and Applied Genetics* 112, 1179-1191.

Carman, J.G., Jamison, M., Elliott, E., Dwivedi, K.K., and Naumova, T.N. (2011). Apospory appears to accelerate onset of meiosis and sexual embryo sac formation in sorghum ovules. *BMC Plant Biology* 11, 9.

Catanach, A.S., Erasmuson, S.K., Podivinsky, E., Jordan, B.R., and Bicknell, R. (2006). Deletion mapping of genetic regions associated with apomixis in *Hieracium*. Proceedings of the National Academy of Sciences of the USA 103, 18650-18655.

Chan, S.W. (2010). Chromosome engineering: power tools for plant genetics. *Trends in Biotechnology* 28, 605-610.

Choi, Y., Gehring, M., Johnson, L., Hannon, M., Harada, J.J., Goldberg, R.B., Jacobsen, S.E., and Fischer, R.L. (2002). Demeter, a DNA glycosylase domain protein, is required for endosperm gene imprinting and seed viability in *Arabidopsis*. *Cell* 110, 33-42.

Copenhaver, G.P., and Preuss, D. (2010). Haploidy with histones. *Nature Biotechnology* 28, 423-424.

d'Erfurth, I., Jolivet, S., Froger, N., Catrice, O., Novatchkova, M., and Mercier, R. (2009). Turning meiosis into mitosis. *PLoS Biol* 7, e1000124.

d'Erfurth, I., Cromer, L., Jolivet, S., Girard, C., Horlow, C., Sun, Y., To, J.P.C., Berchowitz, L.E., Copenhaver, G.P., and Mercier, R. (2010). The CYCLIN-A CYCA1;2/TAM is required for the meiosis I to meiosis II transition and cooperates with OSD1 for the prophase to first meiotic division transition. *PLoS Genetics* 6, e1000989.

Editorial. (2010a). How to feed a hungry world. *Nature* 446, 531-532.

Editorial. (2010b). News feature, food, the growing problem. *Nature* 446, 546-547.

Erilova, A., Brownfield, L., Exner, V., Rosa, M., Twell, D., Scheid, O.M., Hennig, L., and Kohler, C. (2009). Imprinting of the polycomb group gene *MEDEA* serves as a ploidy sensor in *Arabidopsis*. *PLoS Genetics* 5, -.

Farr, C.J., Stevanovic, M., Thomson, E.J., Goodfellow, P.N., and Cooke, H.J. (1992). Telomere-associated chromosome fragmentation - applications in genome manipulation and analysis. *Nature Genetics* 2, 275-282.

Garcia-Aguilar, M., Michaud, C., Leblanc, O., and Grimanelli, D. (2010). Inactivation of a DNA methylation pathway in maize reproductive organs results in apomixis-like phenotypes. *Plant Cell* 22, 3249-3267.

Gilbertson, L. (2003). Cre-lox recombination: Cre-ative tools for plant biotechnology. *Trends in Biotechnology* 21, 550-555.

Grimanelli, D., Leblanc, O., Perotti, E., and Grossniklaus, U. (2001). Developmental genetics of gametophytic apomixis. *Trends in Genetics* 17, 597-604.

Halpin, C. (2005). Gene stacking in transgenic plants--the challenge for 21st century plant biotechnology. *Plant Biotechnology Journal* 3, 141-155.

Han, F., Gao, Z., and Birchler, J.A. (2009). Reactivation of an inactive centromere reveals epigenetic and structural components for centromere specification in maize. *Plant Cell* 21, 1929–1939.

Hare, P. (2008). Research highlights: crop apomixis in the crosshairs. *Nature Biotechnology* 26, 406.

Hecht, V., Vielle-Calzada, J.P., Hartog, M.V., Schmidt, E.D.L., Boutilier, K., Grossniklaus, U., and de Vries, S.C. (2001). The Arabidopsis *Somatic Embryogenesis Receptor Kinase 1* gene is expressed in developing ovules and embryos and enhances embryogenic competence in culture. *Plant Physiology* 127, 803-816.

Horandl, E., and Temsch, E.M. (2009). Introgression of apomixis into sexual species is inhibited by mentor effects and ploidy barriers in the Ranunculus auricomus complex. *Ann Bot-London* 104, 81-89.

Houben, A., and Schubert, I. (2007). Engineered plant minichromosomes: a resurrection of B chromosomes?. *Plant Cell* 19, 2323-2327.

Hwa, C.M., and Yang, X.C. (2008). Fixation of hybrid vigor in rice: opportunities and challenges. *Euphytica* 160, 287-293.

Jones, R.N., Viegas, W., and Houben, A. (2008). A century of b chromosomes in plants: so what? *Ann Bot-London* 101, 767-775.

Jullien, P.E., and Berger, F. (2010). Parental genome dosage imbalance deregulates imprinting in *Arabidopsis*. *PLoS Genetics* 6, e1000885.

Jullien, P.E., Kinoshita, T., Ohad, N., and Berger, F. (2006). Maintenance of DNA methylation during the *Arabidopsis* life cycle is essential for parental imprinting. *The Plant Cell* 18, 1360-1372.

Jullien, P.E., Mosquna, A., Ingouff, M., Sakata, T., Ohad, N., and Berger, F. (2008). Retinoblastoma and its binding partner MSI1control imprinting in *Arabidopsis*. *PLoS Biol.* 6, e194.

Kantama, L., Sharbel, T.F., Schranz, M.E., Mitchell-Olds, T., de Vries, S., and de Jong, H. (2007). Diploid apomicts of the *Boechera holboellii* complex display large-scale chromosome substitutions and aberrant chromosomes. Proceedings of the National Academy of Sciences of the USA 104, 14026-14031.

Kasha, K.J., and Kao, K.N. (1970). High frequency haploid production in barley (*Hordeum vulgare* L.). *Nature* 225, 874-876.

Köhler, C., and Grossniklaus, U. (2005). Seed development and genomic imprinting in plants. In *Progress in molecular and subcellular biology: epigenetics and chromatin*, P. Jeanteur, ed (Berlin-Heidelberg: Springer-Verlag), pp. 237-262.

Krahulcova, A., Krahulec, F., and Rosenbaumova, R. (2010). Expressivity of apomixis in 2n + n hybrids from an apomictic and a sexual parent: insights into variation detected in Pilosella (Asteraceae: Lactuceae). *Sexual Plant Reproduction.*

Leblanc, O., Grimanelli, D., Hernandez-Rodriguez, M., Galindo, P.A., Soriano-Martinez, A.M., and Perotti, E. (2009). Seed development and inheritance studies in apomictic maize-*Tripsacum* hybrids reveal barriers for the transfer of apomixis into sexual crops. *International Journal of Developmental Biology* 53, 585-596.

Li, H., Cao, H., Wang, Y., Pang, Q., Ma, C., and Chen, S. (2009). Proteomic analysis of sugar beet apomictic monosomic addition line M14. *Journal of Proteomics* 73, 297-308.

Luo, M., Bilodeau, P., Koltunow, A., Dennis, E.S., Peacock, W.J., and Chaudhury, A.M. (1999). Genes controlling fertilization-independent seed development in *Arabidopsis thaliana*. Proceedings of the National Academy of Sciences of the USA 96, 296-301.

Martienssen, R.A. (2010). Heterochromatin, small RNA and post-fertilization dysgenesis in allopolyploid and interploid hybrids of *Arabidopsis*. *New Phytologist* 186, 46-53.

Matz, F., Meister, A., and Schubert, I. (2000). An efficient screen for reproductive pathways using mature seeds of monocots and dicots. *Plant Journal* 21, 97-108.

Miernyk, J.A., Pret'ova', A., Olmedilla, A., Klubicova', K., Obert, B., and Hajduch, M. (2010). Using proteomics to study sexual reproduction in angiosperms. *Sexual Plant Reproduction.*

Naqvi, S., Farre, G., Sanahuja, G., Capell, T., Zhu, C., and Christou, P. (2009). When more is better: multigene engineering in plants. *Trends in Plant Science* 15, 48-56.

Nassar, N.M.A., Hashimoto, D.Y., and Ribeiro, D.G. (2010). Genetic, embryonic and anatomical study of an interspecific cassava hybrid. *Genet Mol Res* 9, 532-538.

Nassar, N.M.A., Gomes, P.T.C., Chaib, A.M., Bomfim, N.N., Batista, R.C.D., and Collevatti, R.G. (2009). Cytogenetic and molecular analysis of an apomictic cassava hybrid and its progeny. *Genet Mol Res* 8, 1323-1330.

Ozias-Akins, P., and van Dijk, P.J. (2007). Mendelian genetics of apomixis in plants. *Annual Review of Genetics* 41, 509-537.

Page, D.R., and Grossniklaus, U. (2002). The art and design of genetic screens: *Arabidopsis thaliana*. *Nature Reviews Genetics* 3, 124-136.

Pennisi, E. (2010). Sowing the seeds for the ideal crop. *Science* 327, 802-803.

Phan, B.H., Jin, W.W., Topp, C.N., Zhong, C.X., Jiang, J.M., Dawe, R.K., and Parrott, W.A. (2007). Transformation of rice with long DNA-segments consisting of random genomic DNA or centromere-specific DNA. *Transgenic Research* 16, 341-351.

Pupilli, F., Labombarda, P., Caceres, M.E., Quarin, C.L., and Arcioni, S. (2001). The chromosome segment related to apomixis in *Paspalum simplex* is homoeologous to the telomeric region of the long arm of rice chromosome 12. *Molecular Breeding* 8, 53-61.

Ravi, M., and Chan, S.W.L. (2010). Haploid plants produced by centromere-mediated genome elimination. *Nature* 464, 615-619.

Ravi, M., Marimuthu, M.P.A., and Sidiqqi, I. (2008). Gamete formation without meiosis in *Arabidopsis*. *Nature* 451, 1121-1124.

Sharbel, T.F., Voigt, M.L., Corral, J.M., Thiel, T., Varshney, A., Kumlehn, J., Vogel, H., and Rotter, B. (2009). Molecular signatures of apomictic and sexual ovules in the *Boechera holboellii* complex. *Plant Journal* 58, 870-882.

Sharbel, T.F., Voigt, M.L., Corral, J.M., Galla, G., Kumlehn, J., Klukas, C., Schreiber, F., Vogel, H., and Rotter, B. (2010). Apomictic and sexual ovules of *Boechera* display heterochronic global gene expression patterns. *The Plant Cell* 22, 655-671.

Shukla, V.K., Doyon, Y., Miller, J.C., DeKelver, R.C., Moehle, E.A., Worden, S.E., Mitchell, J.C., Arnold, N.L., Gopalan, S., Meng, X.D., Choi, V.M., Rock, J.M., Wu, Y.Y., Katibah, G.E., Zhifang, G., McCaskill, D., Simpson, M.A., Blakeslee, B., Greenwalt, S.A., Butler, H.J., Hinkley, S.J., Zhang, L., Rebar, E.J., Gregory, P.D., and Urnov, F.D. (2009). Precise genome modification in the crop species *Zea mays* using zinc-finger nucleases. *Nature* 459, 437-U156.

Siddiqi, I., Marimuthu, M.P., and Ravi, M. (2009). Molecular approaches for the fixation of plant hybrid vigor. *Biotechnology Journal* 4, 342-347.

Smith, Z.R., and Long, J.A. (2010). Control of *Arabidopsis* apical-basal embryo polarity by antagonistic transcription factors. *Nature* 464, 423-426.

Sorensen, A.M., Rouse, D.T., Clements, M.A., John, P., and Perotti, E. (2009). Description of a fertilization-independent obligate apomictic species: Corunastylis apostasioides Fitzg. *Sexual Plant Reproduction* 22, 153-165.

Spillane, C., Curtis, M.D., and Grossniklaus, U. (2004). Apomixis technology development-virgin births in farmers' fields? *Nature Biotechnology* 22, 687-691.

Stone, S.L., Braybrook, S.A., Paula, S.L., Kwong, L.W., Meuser, J., Pelletier, J., Hsieh, T.F., Fischer, R.L., Goldberg, R.B., and Harada, J.J. (2008).

Arabidopsis Leafy Cotyledon2 induces maturation traits and auxin activity: implications for somatic embryogenesis. Proceedings of the National Academy of Sciences of the USA 15, 3151–3156.

Su, Y.H., Zhao, X.Y., Liu, Y.B., Zhang, C.L., O'Neill, S.D., and Zhang, X.S. (2009). Auxin-induced *WUS* expression is essential for embryonic stem cell renewal during somatic embryogenesis in *Arabidopsis*. *Plant Journal* 59, 448-460.

Tiwari, S., Schulz, R., Ikeda, Y., Dytham, L., Bravo, J., Mathers, L., Spielman, M., Guzman, P., Oakey, R.J., Kinoshita, T., and Scott, R.J. (2008). *Maternally Expressed PAB C-Terminal*, a novel imprinted gene in *Arabidopsis*, encodes the conserved c-terminal domain of polyadenylate binding proteins. *The Plant Cell* 20, 2387-2398.

Tucker, M.R., and Koltunow, A.M.G. (2009). Sexual and asexual (apomictic) seed development in flowering plants: molecular, morphological and evolutionary relationships. *Functional Plant Biology* 36, 490-504.

van Dijk, P. (2009). Apomixis: basics for non-botanists. In *Lost sex: the evolutionary biology of parthenogenesis*, I. Schon, P. Dijk, and K. Martens, eds (Dordrecht: Springer Netherlands), pp. 47-62.

Vielle-Calzada, J.P., Baskar, R., and Grossniklaus, U. (2000). Delayed activation of the paternal genome during seed development. *Nature* 404, 91-94.

Williams, M. (2010). Grain drain. *Current Biology* 20, R542-R543.

Woodward, F.A., Bardgett, R.D., Raven, J.A., and Hetherington, A.M. (2009). Biological approaches to global environment change mitigation and remediation. *Current Biology* 19, R615–R623.

Yadegari, R., Kinoshita, T., Lotan, O., Cohen, G., Katz, A., Choi, Y., Katz, A.V., Nakashima, K., Harada, J.J., Goldberg, R.B., Fischer, R.L., and Ohad, N. (2000). Mutations in the *FIE* and *MEA* genes that encode interacting polycomb proteins cause parent-of-origin effects on seed development by distinct mechanisms. *Plant Cell* 12, 2367-2381.

Yamada-Akiyama, H., Akiyama, Y., Ebina, M., Xu, Q., Tsuruta, S., Yazaki, J., Kishimoto, N., Kikuchi, S., Takahara, M., Takamizo, T., Sugita, S., and Nakagawa, H. (2009). Analysis of expressed sequence tags in apomictic guineagrass (*Panicum maximum*). *Journal of Plant Physiology* 166, 750-761.

Yao, J.L., Zhou, Y., and Hu, C.G. (2007). Apomixis in Eulaliopsis binata: characterization of reproductive mode and endosperm development. *Sexual Plant Reproduction* 20, 151-158.

Yu, W., Han, F., Gao, Z., Vega, J.M., and Birchler, J.A. (2007). Construction and behavior of engineered minichromosomes in maize. *Proceedings of the National Academy of Sciences of the USA* 104, 8924-8929.

In: Flowering Plants
Editor: Jeremy J. Tellstone

ISBN: 978-1-61324-653-5
© 2011 Nova Science Publishers, Inc.

Chapter 5

SOURCE/SINK RELATIONS IN FRUITING CUTTINGS OF GRAPEVINE (*VITIS VINIFERA* L.) DURING THE INFLORESCENCE DEVELOPMENT

Gaël Lebon, Florence Fontaine, Cédric Jacquard, Christophe Clément and Nathalie Vaillant-Gaveau[*]
Laboratoire de Stress, Défenses et Reproduction des Plantes, URVVC-SE,
EA 2069, Université de Reims Champagne-Ardenne,
UFR Sciences Exactes et Naturelles, Bâtiment 18,
Moulin de la Housse – BP 1039, F-51687 REIMS Cedex 2, France

ABSTRACT

In crops, flower and fruit abscission induce significant yield loss. Dependant on the sink strength of each organ, carbohydrate supply to reproductive structures is a putative physiological cause of this phenomenon. In the present work, the effect of defoliation, as an agent of source/sink modifications, was studied in grapevine (*Vitis vinifera* L.) using fruiting cuttings. In control cuttings only four leaves were allowed to develop. In order to modify artificially source/sink interactions, another set of cuttings was systematically defoliated (0L) whereas the last set was characterized by the free development of all the leaves (AL). The

---

[*] Corresponding author: e-mail : nathalie.vaillant-gaveau@univ-reims.fr

results show that developing leaves have a growth-depressing effect on inflorescences and roots, showing their stronger sink strength. Besides, 0L and AL treatments modify carbohydrate concentrations in all organs of both cvs., especially in inflorescences at crucial steps of male and female organ formation (meiosis). However the variations did not follow the same pathway in each cv., likely due to different carbohydrate metabolism. Whatever, carbohydrate levels in cuttings inflorescences were different than those of vineyard inflorescences.

Abbreviations: CVS: cultivars; GW: Gewurztraminer; PN: Pinot noir

INTRODUCTION

In crops, many environmental conditions and internal mechanisms provoke flower or fruit abscission, leading to important loss of yield (Stopar 1998, Liu et al. 2004). Various putative events may be involved such as nutritional disorders. In this context carbohydrate starvation in the ovary appeared as a key point in the achievement of a functional ovule (Merjanian and Ravaz 1930, Keller and Koblet 1994, 1995, Caspari et al. 1998). In perennial plants insufficient nutrient supply of the flower causing flower and fruit abscission may also be due either to low CO_2 assimilation or to lack of reserves, especially starch (Wardlaw 1990, Rodrigo et al. 2000, Jean and Lapointe 2001). Besides, vegetative and reproductive organs are often competing for the same resources provided by photosynthesis and reserve mobilisation (Wardlaw 1990), meaning that flower and fruit abscission could result from this competition. Indeed, growing inflorescences are not the major sink organs during their development (Ho 1988) because growing roots, leaves and stems attract more efficiently nutrients than inflorescences.

In grapevine (*Vitis vinifera* L.), flower and fruitlet drop so-called 'coulure' is the major cause of yield reduction in several areas (Huglin and Schneider 1998). Each grape cultivar (cv.) exhibits particular coulure sensitivity, meaning that a certain amount of flowers will abort whatever the environmental conditions. For example, the Gewurtzraminer (GW) is highly sensitive, whereas the Pinot noir (PN) is less sensitive to coulure (Huglin and Schneider 1998).

It has been demonstrated that this differential sensitivity is correlated to various amounts of sugars in the flowers during development (Lebon et al. 2004), suggesting that peculiar source/sink interactions in each cv. may contribute to their differential sensitivity to coulure.

Source/Sink Relations in Fruiting Cuttings of Grapevine ...   121

After winter dormancy, the inflorescence development that was initiated and arrested at early stages the previous year is newly stimulated in early spring, synchronously with young roots, annual stems and leaves (Srinivasan and Mullins 1981, Mullins et al. 1992, Boss et al. 2003).

At this moment, leaf photosynthesis is not efficient since leaves are not yet functional. Nutrients are thus exclusively provided by reserves stored in perennial woods, until the leaves reach half of their final size (Alleweldt et al. 1982, Petrie et al. 2000). The transition between the heterotrophic (reserve mobilisation) and the autotrophic (photosynthesis in the leaf) phase usually occurs during flowering, but at different steps of the reproductive effort depending on cvs. (Zapata et al. 2004a, b).

This means that young vegetative organs compete with growing inflorescences for reserves, which may interfere with flower development. Indeed, the ontogenesis of both ovules and anthers is closely correlated with sugars at key steps of the reproductive cell formation (Lebon et al. 2004). In PN, starch is present in the ovule at female meiosis but not in GW, which may contribute to lower coulure sensitivity of the former cv. under stressing environmental conditions.

In vineyard, the experimental approach of source/sink interaction up to flowering is constraining mainly because 1/ flowering occurs only once a year, 2/ it is difficult to collect all the organs for assessments, i.e. the roots can reach up to 10 m underground (Mullins et al. 1992) and 3/ impacts of experimental source/sink perturbations on the quality of flowering can be measured only the year after.

In this context, the fruiting cutting model is of interest because experimental design can be performed all the year (Mullins 1966, Mullins and Rajasekaran 1981, Lebon et al. 2005b). Moreover, this model mimics most of the parameters related to reproduction such as flower, berry or seed sets (Lebon et al. 2005a).

In order to further understand the source/sink interactions in grapevine (*Vitis vinifera* L.) during the development of inflorescences, fruiting cuttings were used under various experimental conditions of defoliation. The growth speed, as well as the sugar content in all the cutting organs were followed along inflorescence development and correlated to the success of the reproduction. The experiments were performed in the GW and the PN cvs owing to their differential sensitivity to coulure.

MATERIALS AND METHODS

Plant Material

Investigations were carried out using the flower abscission sensitive GW and the non-sensitive PN. Dormant cuttings were obtained from canes collected from the INRA vineyard in Bergheim, France (Mullins and Rajasekaran 1981, Lebon et al. 2005b).

Cuttings were collected at the proximal part of the cane and limited to 3 nodes, successively named N0 (proximal), N1 and N2. A key point in the protocol is the sampling of the cane fragments generating the cuttings. The physiology of this fragment, in particular the amount of reserves in the wood is important for obtaining predictable results (Lebon et al. 2005b). Cuttings were treated with cryptonol (2% v/v) to prevent contamination, stored in the dark and forced for at least 2 weeks at 4°C. After 16 h of hydration at 25°-30°C, N0 and N1 were removed and the proximal extremity of the cutting was immerged in indole-3-butyric acid (IBA) at 1 g l^{-1} for 30 s to promote rhizogenesis. This protocol was developed from Mullins (1966) and Mullins and Rajasekaran (1981) and optimized by Lebon et al. (2005b). Cuttings were then planted in 0.5 l pots of perlite:sand (1:1), and transferred to a greenhouse at 20°-30°C (night/day), with the pots placed on a 30°C warming blanket. The photoperiod was 16 h using natural daylight or artificial 400 W "SAUDICLAUDE" lamps supplying $1,000 \text{ } \mu\text{E m}^{-2} \text{ s}^{-1}$ and a relative humidity of 50%. Each pot was watered daily with 100 ml of Coïc and Lesaint (1971) medium.

Source/Sink Treatments

Treatments were carried out until the first inflorescence appeared and aimed at perturbing the source/sink interaction during the flower development. The 4 leaves (4L) treatment was considered as the control according to the results optimisation of the protocol (Lebon et al. 2005b). For this treatment, the four first leaves appearing after the inflorescences developed and the following ones were removed, explaining the absence of leaves at stage 12. In the no leaf (0L) treatment, plants were totally defoliated, so that the inflorescence is the sole sink for cane reserves. In the all leaf (AL) cuttings, all the leaves and vegetative stems were allowed to develop in order to understand the competition between the vegetative and the reproductive aerial organs.

Dry Weight

Inflorescences were collected, from the visible cluster (stage 12) up to the fruit set (stage 27) (Eichhorn and Lorenz 1977). Since meiosis in these cultivars occurs between stages 15 and 17 (Lebon et al. 2005a), this period was subdivided (15+1 day and 15+3 days). At each stage, all the inflorescences, leaves and roots were sampled, frozen in liquid N_2, and stored at -80°C until the determination of sugars. Dry weight (DW) of inflorescences was estimated after freeze-drying for 72 h with a CS5L device (Serail Lyophilisateur®).

Carbohydrate Analysis

Extraction. Inflorescences, roots and leaves were lyophilized 72 h at -80°C with a CS5L device (Serail Lyophilisateur®), and ground in a mortar with Fontainebleau sand and 10 volumes of ethanol 80°. Sugars were then extracted for 15 min at 84°C under continual agitation. After adjusting the volume to 5 ml with distilled water, the extract was centrifuged at 4°C for 10 min at 11,000 g. The supernatant was used for soluble sugar determination. For starch, the pellet was suspended in a mixture containing dimethylsulfoxide (DMSO):hydrochloric acid 8 N (8:2) and starch was dissolved during 30 min at 60°C under continual agitation. After cooling, the extract was centrifuged at 20°C for 10 min at 13,000 g and the supernatant was kept at -80°C until use.

Sucrose, glucose and fructose assay (Bergmeyer 1974). Sucrose, glucose and fructose were assayed because they are the major sugars in grapevine (Glad et al. 1992). D-glucose was phosphorylated and oxidized in the presence of NADP to gluconate-6-phosphate and NADPH, H^+. The amount of NADPH, H^+ formed was determined by means of its absorbance at 340 nm. Fructose was phosphorylated to fructose-6-phosphate by a hexokinase in the presence of ATP. Fructose-6-phosphate was then converted to glucose-6-phosphate by a phosphoglucoisomerase. Glucose-6-phosphate formed was tested as described above and a blank was performed without phosphoglucose isomerase. Sucrose was hydrolyzed to D-glucose and D-fructose in the presence of a β-fructosidase. D-glucose formed was then determined as described above and compared with a blank without β-fructosidase.

Starch assay. Aliquots of 100 µl of the extract were used to determine starch concentration. The aliquot was mixed with 100 µl of Lugol solution (0.03% I_2 and 0.06% KI in diluted 0.05 N HCl). After 15 min, the absorbance was read spectrophotometrically at 600 nm. A blank was performed with the starch solvent (DMSO:HCl, 8:2) instead of the extract.

Statistical Analysis

At least five assays were performed for each stage of flower development, and three independent readings were carried out for each extract. For carbohydrate determination, results are expressed in % dry weight ± SE. Statistical analyses were carried out using a Student's t test. A 2% probability was considered significant.

RESULTS

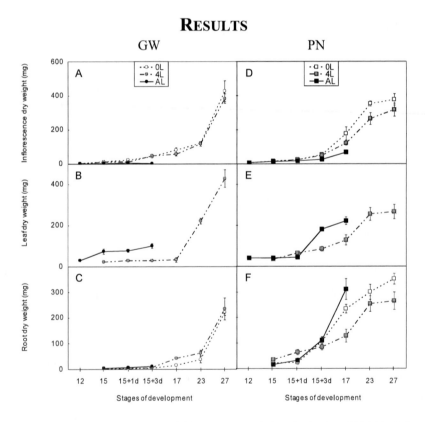

Figure 1. Dry Weight variations in different organs of GW (A,B,C) and PN (D,E,F) fruiting cuttings during their development: (A-D) inflorescences, (B-E) leaves and (C-F) roots. Measures are stopped when plants are AL-treated due to inflorescence necrosis. Values are means (n=5) ± SE.

In the control (4L) and 0L cuttings, the dry weight of inflorescences followed the same pathway in both cvs. (Figs. 1A, D). It was characterized by an exponential increase from stage 15+3d up to fruit set (stage 27). In AL treated cuttings, the inflorescences exhibited a significantly weaker growth. Later on, in both cvs., the AL treatment induced necrosis of the inflorescence, precisely at stages 15+3d in GW and 17 in PN inflorescences (Figs. 1A, D).

Leaves appeared on cuttings at different moments depending on the treatment. In AL cuttings, leaves appeared at stage 12 whereas they emerged only at stage 15 in 4L-treated plants (Figs. 1B, E). At stage 15+3 days, AL cuttings leaves of GW and PN respectively weighted 3.4-fold and 2.2-fold more than 4L ones. Considering the three treatments, roots appeared on the fruiting cutting from stage 15 in both cvs. (Figs.1C, F). In GW root development was synchronous in the three treatments, whereas in PN, root growth was significantly favoured in AL and 0L cuttings when compared to 4L ones.

Carbohydrate Contents

Inflorescences. In GW inflorescences of 4L cuttings starch concentrations decreased from 2.48±0.17 to 0.31±0.14% DW from stage 12 to stage 15 and remained statistically constant during the end of flower development (Fig. 2A). In AL cuttings the inflorescence exhibited the highest amount of starch that was stable from stage 12 until stage 15+1d. Afterwards, it decreased dramatically to 0.9±0.19% DW at stage 15+3d (Fig. 2A) and the inflorescence thus degenerated. In 4L and 0L cuttings, the inflorescences had the same starch concentration from stage 15 up to fruit set. In these cuttings, soluble carbohydrate concentrations followed the same variations (Figs. 2B, C, D), though sucrose and fructose concentrations were higher in 4L treated cuttings. In the inflorescences of AL cuttings, sucrose concentration was lower than in other treatments, whereas glucose concentration was higher at stages 15+1d and 15+3d and fructose at stage 15 (Figs. 2B, C).

Whatever the treatment, starch variations in the inflorescences of PN had the same profile and fluctuated between 0.19±0.03 and 0.5±0.15% DW, except in 4L-treated inflorescences at stage 12 (Fig. 2E). Similarly, fluctuations of sucrose concentrations were similar whatever the treatment (Fig. 2F): after a significant decrease from stage 12 to stage 15+1d, sucrose concentration slowly dropped down and reached very low values at fruit set (stage 27). The three treatments induced the same pattern in glucose and fructose variations in inflorescences with a decrease from the stages 12 to 15, except for the glucose in inflorescences of AL cuttings at stage 12 (Figs. 2G, H). Thereafter, hexose levels in inflorescences of 4L and AL cuttings were constant around 1% DW until fruit set although it decreased regularly down to 0.33±0.01% DW in 0L cuttings. Moreover, hexoses were significantly more abundant in inflorescences of 4L- than in those of 0L-cuttings.

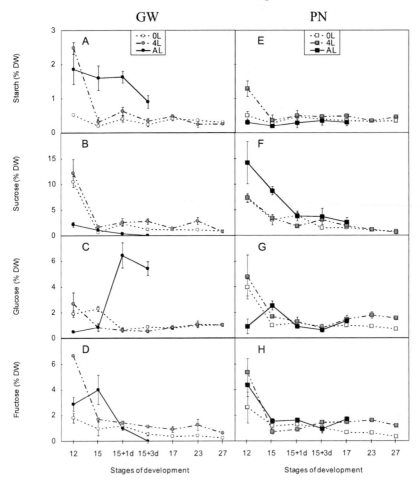

Figure 2. Changes in some carbohydrate levels in inflorescences of GW (A,B,C,D) and PN (E,F,G,H) fruiting cuttings during their development: (A-E) starch, (B-F) sucrose, (C-G) glucose and (D-H) fructose. Measures are stopped when plants are AL-treated due to treated due to inflorescence necrosis. Values are means (n=5) ± SE.

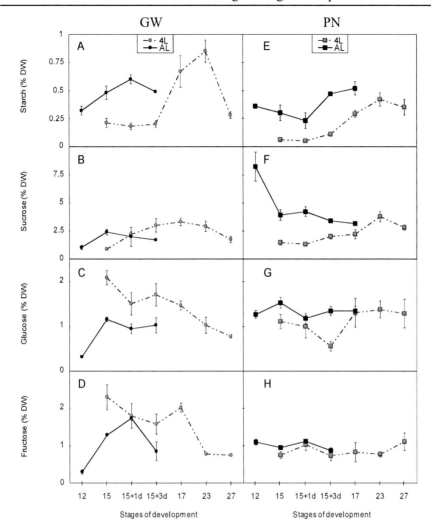

Figure 3. Changes in some carbohydrate levels in leaves of GW (A,B,C,D) and PN (E,F,G,H) fruiting cuttings during their development: (A-E) starch, (B-F) sucrose, (C-G) glucose and (D-H) fructose. Measures are stopped when plants are AL-treated due to treated due to inflorescence necrosis. Values are means (n=5) ± SE.

Leaves. Considering leaves, no results were obtained in the 0L treatment since leaves were systematically removed. The AL treatment induced higher starch levels in GW leaves than the 4L treatment (Fig. 3A). However, in this cv. sucrose concentrated in leaves of 4L cuttings after stage 15+1d (Fig. 3B), as well as glucose and fructose (Figs. 3C, D). In PN, the AL treatment generated a higher accumulation of carbohydrates (at least starch, sucrose and glucose) when compared to the 4L treatment (Figs. 3E, F, G, H).

Roots. In the roots of cuttings, carbohydrate concentrations decreased from stage 15 to stage 27 whatever the treatment and the cultivar (Fig. 4). In roots of 4L GW cuttings, starch level decreased rapidly from stages 15 to 17, and thereafter remained constant at 0.15% DW (Fig. 4A). In the 0L cuttings starch concentration in roots is constant during the whole of floral development (Fig. 4A).

Regarding sucrose, the 0L treatment induced the highest concentrations, whereas the 4L treatment mostly reduced the sucrose level during the whole inflorescence development (Fig. 4B). Considering glucose, roots of 0L cuttings exhibited the lowest glucose concentration from stage 15 to stage 27 (Fig. 4C). In the meantime, roots of 4L and AL cuttings accumulated glucose respectively from 1.5- to 4.8-fold and from 1.8- to 4.2-fold more than the roots of 0L cuttings. Fructose content in roots of AL cuttings dropped rapidly between stages 15 and 15+3d from 10±0.68 to 1.98±0.35% DW (Fig. 4D). 0L and 4L treatments provoked a slight fructose decrease to reach 0.35±0.06 and 0.45±0.03% DW, respectively.

All the treatments provoked similar responses in the PN cuttings. In roots of 4L cuttings, starch level increased from stages 15 to 23 (from 0.06±0.02 to 0.42±0.06% DW), and remained constant until fruit set (Fig. 4E). The profile of starch levels in roots of AL and 0L cuttings was opposite. Indeed, in roots of 0L cuttings starch contents decreased slowly during inflorescence development.

In roots of AL cuttings, starch content decreased by 2.9 between stages 15 and 15+1d, and thereafter had the same values than roots of 0L cuttings. As in GW roots, the 0L treatment induced the highest sucrose concentration and the 4L-one the lowest one (Fig. 4F). In the case of glucose, the AL treatment induced a constant decrease of glucose concentrations in roots from stage 15 to 17 (Fig. 4G), whereas the two other treatments provoked globally the same pattern in glucose levels. Finally, the AL treatment induced a 5.4-fold decrease of fructose levels in roots from stage 15 to 17 to reach 0.92±0.14% DW (Fig. 4H). Although the fluctuation of fructose content in roots was similar in 0L and 4L cuttings, levels of fructose were significantly higher in roots of 4L cuttings.

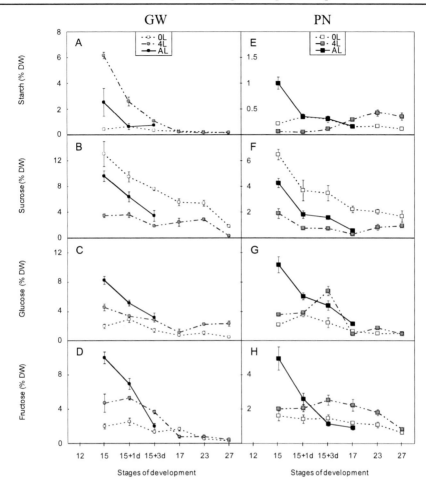

Figure 4. Changes in some carbohydrate levels in roots of GW (A,B,C,D) and PN (E,F,G,H) fruiting cuttings during their development: (A-E) starch, (B-F) sucrose, (C-G) glucose and (D-H) fructose. Measures are stopped when plants are AL-treated due to treated due to inflorescence necrosis. Values are means (n=5) ± SE.

DISCUSSION AND CONCLUSION

In fruiting cuttings of grapevine, there is a competition between the vegetative and the reproductive organs for the carbohydrates that can be mobilised from the woody perennial tissues of the plant. In this artificial model, the vegetative stems, leaves and roots are stronger sinks than the inflorescences which are deprived of sugars and thus degenerate when vegetative stem growth is not experimentally limited.

In both cultivars, during the first stages of flower development, leaves and stems have a growth-depressing effect on inflorescences and induce carbohydrate retention of these organs as observed previously (Mullins 1967, 1968, Mullins and Rajasekaran 1981). As in other species, grapevine stems, inflorescences and roots compete for nutrient supply. In spring during bud burst, CO_2 assimilation in the leaves is not sufficient to allow their development explaining that leaves represent a strong sink for photoassimilates rather than a source (Wardlaw 1990, Jean and Lapointe 2001). In the meantime, emerging inflorescences have a limited ability to attract photoassimilates when compared to stem development (Mullins et al. 1992). In our study, the sink strength of the stem in the AL cuttings is revealed by the simultaneous growth of leaves and the low growth followed by the necrosis of inflorescences. This may be explained by the continuous need of sugar by developing new leaves in the AL cuttings, whereas in the 4L ones, the stem growth and the subsequent carbohydrates request is limited by the removal of any new leaf or shoot meristems. Moreover, the constant decrease of root carbohydrate levels at the benefit of leaves exhibiting constant or increasing sugar contents further illustrates this phenomenon. Indeed, as in tomato, the relative sink strength of leaf and shoot apex is higher than the root one (Shishido et al. 1999).

The various treatments modify carbohydrate contents of vegetative (leaf and roots) and reproductive (inflorescences) organs in both cvs. In both GW and PN, the content of sugars in the inflorescence globally increased (especially fructose) following the 4L-treatment, whereas in the inflorescence of the 0L cuttings, the carbohydrate level is not influenced. In papaya, a 60% defoliation induces the decrease of fruit carbohydrate contents (Zhou et al. 2000). Moreover, the enzymes involved in carbohydrate metabolism are known to have differential activities in the inflorescences of the both cvs, explaining that sugars are accumulated in different forms.

In both tested cvs. carbohydrates accumulate in the leaves of AL cuttings when compared to 4L ones, mainly soluble carbohydrates in PN and starch in GW. This accumulation of sugars in leaves is in contradiction with previous results obtained in cherry (Layne and Flore 1993): 75% defoliation provokes an increase of leaf sucrose and starch due to a stimulation of CO_2 assimilation. The accumulation of sugar may thus be the consequence of a higher CO_2 assimilation or/and a stronger reserve mobilisation.

In cutting roots of both cvs., the 0L treatment modifies carbohydrate levels, as revealed by an increase of sucrose content and a synchronous decrease of glucose and fructose contents. These modifications are due to the major source status of roots (Loescher et al. 1990, Bates et al. 2002, Zapata et al. 2002a), and to their minor sink importance (Candolfi-Vasconcelos et al. 1994, Shishido et al. 1999). The source status of roots implies that they intensely synthesize sucrose, which is the major form of carbohydrate transport. Thus, the variations of sucrose levels are the consequence of some carbohydrate metabolism related enzymes, such as the reduction of invertase and the increase of sucrose synthase activities (Weschke et al. 2003, Koch 2004). Indeed, we have previously demonstrated that the fluctuations of sucrose content in inflorescences correspond to the variations of cytosolic and wall-bound invertases, as well as sucrose synthase (synthesis) activities.

Moreover, roots become the only carbohydrate source when leaves are removed (0L treatment), explaining their decrease of starch content, as observed in young and old stolons of white clover (Gallagher et al. 1997). Furthermore, root carbohydrates are reduced during the inflorescence development whatever the treatment. This was already reported in fruiting cuttings (Bartolini et al. 1996) and may be due to carbohydrate needs of developing vegetative and reproductive organs (Jean and Lapointe 2001).

Carbohydrate variations in vegetative and reproductive organs are different between GW and PN. GW roots and leaves accumulate higher quantities of starch and fructose than PN ones. Different patterns of carbohydrate distribution are ever found in inflorescences between these cultivars (Lebon et al. 2004). And these cultivar distinctions are due to different carbohydrate metabolism (Lebon et al. 2005a). Thus, GW fruiting cuttings principally favour carbohydrate accumulation in roots and leaves, and significantly more than PN. Preferentially storing carbohydrate reserves in vegetative organs rather than in inflorescences could explain the higher sensitivity of GW to flower abscission. Carbohydrate levels of inflorescences are crucial for flower fate (Rodrigo et al. 2000, Jean and Lapointe 2001, Ruiz et al. 2001, Iglesias et al. 2003) and GW inflorescences contain low carbohydrate levels during development. Therefore, it is likely that GW inflorescences have not enough available sugars to counteract environmental stresses, especially when they occur during meiosis, explaining their sensitivity to coulure.

In cuttings as in vineyard, GW and PN inflorescences exhibit major differences of carbohydrate levels during the crucial steps in male and female organ formation, i.e., at meiosis. In both cvs., female meiosis occurs from stages 15 to 17, and cvs. exhibit different pattern of carbohydrate distribution during this period: in PN, developing inflorescences mainly accumulate hexoses, whereas in GW ones starch and sucrose are stored (Lebon et al. 2004). The higher pool of available hexoses in PN developing inflorescences may explain the lower sensitivity to flower abscission in PN under stress conditions (Lebon et al. 2004, Zapata et al. 2004b). Moreover, this critical period is marked by the transition between mobilization of starch from the root reserves and the translocation of photosynthates from fully expanded leaves (Zapata et al., 2004a), and coincides with variations of carbohydrates in inflorescences (Lebon et al., 2005a). However, although carbohydrate level differences exist in GW and PN inflorescences, these variations during flower development of fruiting cuttings differ from those observed in vineyard. Indeed, contrarily to vineyard, sucrose is more accumulated in PN inflorescences than in GW ones, and other carbohydrate levels are identically concentrated in both cultivars. These differences between fruiting cutting and vineyard are likely caused by different carbohydrate metabolism due to the absence of root during the first steps of the inflorescence development. Indeed, during this period, carbohydrate supply only originates from wood reserves. Although the development of reproductive structures is identical in fruiting cutting and vineyard (Lebon et al. 2005b), the fluctuation of carbohydrate levels differs between these two models.

ACKNOWLEDGMENTS

The authors thank C. Guèze for technical assistance and acknowledge ONIVINS and the Mumm-Perrier-Jouët company for financial supports.

REFERENCES

Alleweldt G, Eibach R, Rühl E (1982). Investigations on gas exchange in grapevine. I. Influence of temperature, leaf age and daytime on net photosynthesis and transpiration. *Vitis* 21: 93-100.

Bartolini G, Pestelli P, Topni MA, Di Monte G (1996). Rooting and carbohydrates avaibility in *Vitis* 140 Ruggeri stem cuttings. *Vitis* 35(1): 11-14.

Bates TR, Dunst RM, Joy P (2002). Seasonal dry matter, starch, and nutrient distribution in 'Concord' grapevine roots. *HortScience* 37(2): 313-316.

Bergmeyer HU (1974). *Methods of enzymatic analysis*. Verlag Chemie, Weinheim Basel.

Boss PK, Buckeridge EJ, Poole A, Thomas MR (2003). New insights into grapevine flowering. *Funct Plant Biol* 30: 593-606.

Candolfi-Vasconcelos MC, Candolfi MP, Koblet W (1994). Retranslocation of carbon reserves from the woody storage tissues into the fruit as a response to defoliation stress during the ripening period in *Vitis vinifera* L. *Planta* 192: 567-573.

Caspari HW, Lang A, Alspach P (1998). Effects of girdling and leaf removal on fruit set and vegetative growth in grape. *Am J Enol Vitic* 49: 359-366.

Coïc F, Lesaint L (1971). Comment assurer une bonne nutrition en eau et en ions minéraux en horticulture? *Hortic Française* 8: 11-14.

Eichhorn KW, Lorenz DH (1977). *Phöenologische Entwicklungsstadie. Der rebe. Nachrichtenb.* Deutsch Pflanzenschutzd (Braunschweig) 29: 119-120.

Gallagher JA, Volenec JJ, Turner LB, Pollock CJ (1997). Starch hydrolytic enzyme activities following defoliation of white clover. *Crop Sci* 37: 1812-1818.

Glad C, Regnard JL, Querou Y, Brun O, Morot-Gaudry JF (1992). Flux and chemical composition of xylem exudates from Chardonnay grapevines: temporal evolution and effect of recut. *Am J Enol Vitic* 43: 275-282.

Ho LC (1988). Metabolism and compartmentation of imported sugars in sink organs in relation to sink strength. *Annu Rev Plant Physiol* 39: 355-378.

Huglin P, Schneider C (1998). Biologie et écologie de la vigne. *Lavoisier Tech.* Doc. 2nd edition, 372 pp.

Iglesias DJ, Tadeo FR, Primo-Millo E, Talon M (2003). Fruit set dependence on carbohydrate availability in *Citrus* trees. *Tree Physiol* 23: 199-204.

Jean D, Lapointe L (2001). Limited carbohydrate availability as a potential cause of fruit abortion in *Rubus chamaemorus. Physiol Plant* 112: 379-387.

Keller M, Koblet W (1994). Is carbon starvation rather than excessive nitrogen supply the cause of inflorescence necrosis in *Vitis vinifera* L.? *Vitis* 33: 81-86.

Keller M, Koblet W (1995). Dry matter and leaf area partitioning, bud fertility and second season growth of *Vitis vinifera* L.: responses to nitrogen supply and limiting irradiance. *Vitis* 34: 77-83.

Koch KE (2004). Sucrose metabolism: regulatory mechanisms and pivotal roles in sugar sensing and plant development. *Curr Opin Plant Biol* 7: 235-246.

Layne DR, Flore JA (1993). Physiological responses of *Prunus cerasus* to whole-plant source manipulation. Leaf gas exchange, chlorophyll fluorescence, water relations and carbohydrate concentrations. *Physiol Plant* 88: 44-51.

Lebon G, Brun O, Magné C, Clément C (2004). Flower abscission and inflorescence carbohydrates in sensitive and non-sensitive cultivars of grapevine. *Sex Plant Reprod* 17: 71-79.

Lebon G, Brun O, Magné C, Clément C (2005a). Photosynthesis of the grape (*Vitis vinifera*) inflorescence. *Tree Physiol* 25: 633-639.

Lebon G, Duchêne E, Brun O, Clément C (2005b). Phenology of flowering and starch accumulation in grape (*Vitis vinifera* L.) cuttings and vines. *Ann Bot* (Lond) 95: 943-948.

Liu F, Jensen CR, Andersen MN (2004). Pod set related to photosynthetic rate and endogenous ABA in soybeans subjected to different water regimes and exogenous ABA and BA at early reproductive stages. *Ann Bot* (Lond) 94: 405-411.

Loescher WH, McCamant T, Keller JD (1990). Carbohydrate reserves, translocation, and storage in woody plant roots. *HortScience* 25(3): 274-281.

Merjanian A, Ravaz V (1930). Sur la coulure de la vigne. *Prog Agri Vitic* 49: 545-550.

Mullins MG (1966). Test plants for investigations of the physiology of fruiting in *Vitis vinifera* L. *Nature* 209: 419-420.

Mullins MG (1967). Morphogenetic effects of roots and of some synthetic cytokinins in *Vitis vinifera* L. *J Exp Bot* 18(55): 206-214.

Mullins MG (1968). Regulation of inflorescence growth in cuttings of the grape vine (*Vitis vinifera* L.). *J Exp Bot* 19(60): 532-543.

Mullins MG, Bouquet A, Williams LE (1992). *Biology of the grapevine.* Mullins (Ed), University Press, Cambridge, UK, 239 pp.

Mullins MG, Rajasekaran K (1981). Fruiting cuttings: revised method for producing test-plants of grapevine cultivars. Am J Enol Vitic 32(1): 35-40.

Petrie PR, Trought MCT, Howell GS (2000). Influence of leaf ageing, leaf area and crop load on photosynthesis, stomatal conductance and senescence of grapevines (*Vitis vinifera* L. cv. Pinot noir) leaves. *Vitis* 39: 31-36.

Rodrigo J, Hormaza JI, Herrero M (2000). Ovary starch reserves and flower development in apricot (*Prunus armeniaca*). *Physiol Plant* 108: 35-41.

Ruiz R, Garcia-Luis A, Honerri C, Guardiola JL (2001). Carbohydrate availability in relation to fruitlet abscission in *Citrus*. *Ann Bot* 87: 805-812.

Shishido Y, Kumakura H, Nishizawa T (1999). Carbon balance of a whole tomato plant and the contribution of source leaves to sink growth using the $^{14}CO_2$ steady-state feeding method. *Physiol Plant* 106: 402-408.

Srinivasan C, Mullins MG (1981). Physiology of flowering in the grapevine. A review. *Am J Enol Vitic* 32(1): 47-63.

Stopar M (1998). Apple fruitlet thinning and photosynthate supply. *J Hortic Sci Biotechnol* 73(4): 461-466.

Wardlaw IF (1990). The control of carbon partitioning in plants. *New Phytol* 116: 341-381.

Weschke W, Panitz R, Gubatz S, Wang Q, Radchuk R, Weber H, Wobus U (2003). The role of invertases and hexose transporters in controlling sugar ratio in maternal and filial tissues of barley caryopses during early development. *Plant J* 33: 395-411.

Zapata C, Deléens E, Chaillou S, Magné C (2004a). Partitioning and mobilization of starch and N reserves in grapevine (*Vitis vinifera* L.). *J Plant Physiol* 161: 1031-1040.

Zapata, C, Déléens E, Chaillou S, Magné C (2004b). Mobilisation and distribution of starch and total N in two grapevine cultivars differing in their susceptibility to shedding. *Funct Plant Biol* 31: 1127-1135.

Zhou L, Christopher DA, Paull RE (2000). Defoliation and fruit removal effects on papaya fruit production, sugar accumulation, and sucrose metabolism. *J Amer Soc Hort Sci*, 125(5): 644-652.

In: Flowering Plants
Editor: Jeremy J. Tellstone

ISBN: 978-1-61324-653-5
© 2011 Nova Science Publishers, Inc.

Chapter 6

IMPROVEMENT STRATEGIES TO CONTROL ARCHITECTURE AND FLOWERING IN ORNAMENTAL PLANTS SUCH AS AZALEA[*]

M. Meijón, M. J. Cañal, R. Rodríguez, and I. Feito
Oviedo University, School of Biology, Oviedo , Spain

INTRODUCTION

The evergreen azalea (*Rhododendron* L. sp) is a woody shrub widely used in gardens and in fact the genus *Rhododendron* is among the most popular landscape plants in Europe and North America. It is also sold as a greenhouse-grown potted plant, marketed in flower, for decorative indoor use. The genus contains approximately 1000 described species and thousands of commercial hybrids with new cultivars entering the market every year. Plants are grown as densely branched shrubs in containers of various sizes. However, many cultivars show a very strong growth tendency and an irregular shape.

The ornamental industry's difficulties in producing compact and well branched plants with high floral quality have been the subject of extensive study, the problems being compounded by the fact that each species, and even each cultivar, requires a specific protocol.

---

[*]A version of this chapter was also published in Ornamental Plants: Types, Cultivation and Nutrition, edited by Joshua C. Aquino, published by Nova Science Publishers, Inc. It was submitted for appropriate modifications in an effort to encourage wider dissemination of research.

Azalea production has developed considerably since 1989 (Marosz and Matysiak, 2005). One of the main growth problems in these species is excessive shoot growth during container production, and often more than three years (six to eight growth flushes) is needed following propagation for full flowering to be achieved (Marosz and Matysiak, 2005). Thus, to improve retail nursery sales of rhododendrons, especially azalea, shoot length needs to be controlled and flowering advanced.

In production of ornamental crops, high fertility and an unlimited supply of water are used to enhance plant growth. This often results in a plant that is taller than desired. Photoperiod manipulation is used to control flower induction in some species though increased stem elongation can be a consequence when artificial light is used to produce a long photoperiod. In addition, the warm temperatures required to get a crop to flower on schedule can cause plants to elongate, regardless of photoperiod. Growth retardants can be used in each of these situations to achieve an ideal height and in contrast to in agronomy, where growth retardants are only applied to few species, in ornamental horticulture these compounds are used on many species (Basra, 2000).

Growth retardants have been used in the commercial production of potted rhododendrons for over 40 years (Marosz and Matysiak, 2005). Also a good deal of research has been conducted into the effect of growth regulators on controlling shoot elongation and flower bud formation in woody plants, especially in *Rhododendron* sp (Gent, 1995; Marosz and Matysiak, 2005; Meijón et al., 2009).

In this chapter the efficiency of different strategies to improve ornamental plant quality is analyzed with the additional aim of lowering costs in azalea production. In addition, quantitative methods to assess the effectiveness of these treatments (such as image analysis or classic biometry) are reviewed.

STRATEGIES TO IMPROVE ORNAMENTAL PLANT QUALITY

Despite the increase in the demand for ornamental plants in recent years, there is no clear international consensus on the criteria that define plant quality, nor do the countries with greater involvement in the ornamental sector have their own well-established policy. In the UK, research and development projects have been underway since 2002 to establish and implement parameters to compare plants produced by different systems (Edmondson and

Parsons, 2002; 2005). However the basic quality criteria which plant nurseries aim for are:

- Disease and pest free plants.
- Plants without defects caused by weather, pests, diseases or deficient nutrition.
- Plants with a good nutritional status that provides a good/attractive leaf color.
- Good flowering in both quantity and quality.
- Plant architecture that is rounded and compact with good branching from the base.

Given that numerous cultivars of azalea show a very strong growth tendency and an irregular shape, the ability to control shoot elongation at the same time as improving floral quality are key factors for planning production in this ornamental species.

The control of growth and floral development can be mediated through nutrient input. Moderate phosphorus or nitrogen stress can reduce vegetative growth and produce compact plants (Sharma et al., 1979), while the application of fertilizers rich in phosphorus can advance and enhance flowering (Skinner and Matthews, 1989). Other techniques extensively used in ornamental plant production to improve architecture and flowering are those based on the control of environmental factors such as photoperiod (quality and duration) (Islam et al., 2005), temperature (Moe and Heins, 1990) or water stress (Fernández et al., 2010). However to reach the required balance for each situation and species is complicated in all these techniques.

In addition to manipulation of nutrient input, many nurserymen use techniques such as manual shoot-pinching to reduce apical dominance and encourage the outgrowth of side-shoots. The major disadvantage of such manual pinching is the high labor input, particularly since pinching may have to be repeated several times due to the reestablishment of a single dominant shoot. Alternative methods of pinching using various chemical agents, such as plant growth regulators (PGRs) (Halmann, 1990; Basra, 2000) or fatty acid methyl esters (Shu et al., 1981), offer the opportunity of reducing labor costs although the response of different species in terms of shoot growth and flowering is highly variable and not all chemicals guarantee effective control (Meijón et al., 2009).

The majority of plant growth regulators employed in ornamental plant culture are chemical growth retardants used, fundamentally, to control the size of plants, improve compactness and enhance flowering (Halmann, 1990; Basra, 2000; Marosz and Matysiak, 2005). In addition, these compounds also increase other functional aspects, such as the ability to resist the negative effects of water stress (Navarro et al., 2007) or low temperatures in winter (Fletcher et al., 2000), beside improving the color of the leaves and chlorophyll content. A chemical plant growth regulator is a natural or synthetic chemical substance, usually organic, that in very small quantities regulates or controls some aspects of plant growth, e.g. stem length, flowering, leaf abscission or winter hardiness (Basra, 2000). A great disadvantage of using these chemicals is the development of phytotoxic symptoms, such as chlorosis, deformed leaves or damaged flowers, which may persist for a long time (Gent, 1995, 1997, 2004). Therefore, it is important to establish specific protocols for different species under particular environmental conditions (Smit et al., 2005).

One such method of chemical pinching is the application of fatty acid methyl esters from C6 to C18. Lower alkyl esters of C8–C12 fatty acids and C8–C10 fatty alcohols have been found to selectively kill the terminal meristem of a wide variety of plants without damaging the axillary meristems, foliage or stem tissue (Shu et al., 1981). In azalea however, the application of fatty acids has been shown to be inadequate since the effect on shoot elongation is insufficient and multiple foliar sprays are required, which increases labor requirements and costs (Meijón et al., 2009). Furthermore, its destructive impact on the apices is reflected in the further development of the unaffected branches leading to a loss of homogeneity (Richards and Wilkinson, 1984), in addition to it disrupting leaf development (leading to deformity and loss of aesthetic quality) and finally reducing the density of the flower as a consequence of the destruction of flower buds (Richards and Wilkinson, 1984; Meijón et al., 2009)

USE OF NUTRITIONAL BALANCE AND WATER CONDITIONS TO IMPROVE ARCHITECTURE AND FLOWERING

For optimum plant growth, it is essential that fertilizers provide sufficient nutrients for initial growth, followed by a uniform supply that synchronizes well with the nutrient requirement of the crop (Sharma, 1979).

By modifying the balance of nitrogen (N), phosphorus (P) and potassium (K) at the times in the growth process when these elements are more crucial it is possible to improve both flowering and vegetative growth. In general, nitrogen is the element that most often limits plant growth, while phosphorus and potassium are key elements for floral development.

Ristvey et al., (2007) and Hansen and Lynch (1998) observed that, in azalea, increasing N fertilization rate promoted shoot growth, whereas decreasing N and P fertilization rate promoted root growth and increased uptake efficiency. These results suggest that the high P concentrations used in many horticultural systems in order to advance and enhance flowering (Skinner and Matthews, 1989) may have no benefit in terms of shoot growth and may actually be detrimental to root growth. Conversely, inadequate phosphorus or potassium in the soil may suppress flower bud production. However it may be the case that each species or hybrid requires a specific individual formula.

In azalea production, plant nurseries use NPK balances which vary with the season, presumably based on their assumptions of the plants' requirements at each stage of development. However, the same fertilization is usually applied to all cultivars even though it has been found that development patterns differ considerably (Meijon et al., 2009).

Seasonal fertilization is based on a balance of nutrients that enhances the proportion of phosphorus during flowering induction, that of potassium during the development of flower buds and to facilitate the hardening off process in late summer early fall, and that of nitrogen in spring in order to improve vegetative growth. Results from trials with Blaauw's Pink and Johanna cultivars of azalea showed that the success of fertilization depends largely on the development pattern of each cultivar. Thus, Johanna, which presents low vegetative growth (Meijón et al., 2009), achieved the most appropriate development with fertilization composed of lower doses of phosphorus and potassium and maximum ammonium ratio. On the other hand, in Blaauw's Pink, which shows high growth capability (Meijón e al, 2009), no appropriate development pattern under any of the tested fertilization conditions was found, although there were slight improvements when the proportion of phosphorus and potassium in the nutrition solution was increased and that of the ammonium was reduced.

At the same time there are many reports providing evidence of water deficit promoting floral initiation, although few studies have demonstrated that it enhances flowering in many important horticultural and forestry species (Stern et al., 2003; Sharp et al., 2009). Cameron et al. (1999) reported that plant water deficit enhanced flower bud formation in Rhododendron, but this only occurred when treatments were applied after floral initiation. In Blaauw's

Pink and Johanna cultivars of azalea moderate water stress (20-30 bar) was found to improve floral development although there were difficulties maintaining adequate water pressure. Water stress influences hormonal balance, especially of cytokinins and abscisic acid, which are known to affect bud differentiation (Cobersier et al., 2003; Stern et al., 2003, Meijón et al., 2010a unpublished)

TEMPERATURE AND PHOTOPERIOD IN CONTROLLING DEVELOPMENT OF AZALEA

Among other factors, light, temperature, and the synergistic effect of both environmental cues are floral inductors in numerous species (Wilkie et al., 2008). Consequently, interventions in temperature and photoperiod are common practices in hortofloricultura to promote adjustment of production to meet market demands.

Change in temperature can control vegetative growth as well as flowering and fruiting processes. The accumulation of chilling hours and growing degree day (GDD), are indicators used, primarily in the agronomic sector, to predict certain behaviors of plants.

GDD is used to monitor the effect of temperature during the season on crop growth, development and maturity. It is calculated by subtracting a base temperature (assumed to be the temperature at which plant development slows or stops) from the daily average temperature e.g. if the average daily temperature is 15°C and the base temperature is 10°C there are 5 GDD's. If the number is less than 0, the GDD is taken as 0. GDD's for each day are usually added together to give a GDD accumulation. This accumulation can be used as a comparison with previous years or to estimate the time for a crop to reach maturity. In tests carried out in controlled conditions in Blaauw's Pink and Johanna cultivars of azalea it has been found that a minimum of 1300 GDD units under long day photoperiod are necessary for bloom to take place (Meijón et al., 2010b unpublished).

An additional technique for the regulation of plant architecture is control of temperature. Daily alterations in day (DT) and night temperature (NT) affect stem and internode elongation in most plants. In general, plants grow tall with long internodes when the DT is lower than the NT. These thermoperiodic responses are utilized to control plant height in commercial plant production in climate-controlled conditions (Grindal et al., 1998).

With respect to light, photoperiod change and the quality of the light source are essential tools in current nursery production of ornamental plants because daylength is a reliable indicator of the time of year, enabling developmental events to be scheduled to coincide with particular environmental conditions. Photoperiod control activates signals that promote numerous processes during plant development. Among the responses thought likely to be regulated by photoperiod are: flowering, tuber formation, onset of dormancy, cambial activity and leaf abscission (Jackson, 2009).

The azalea cultivars Blaauw's Pink and Johanna require a long photoperiod with minimum hours of light and a minimum GDD (see above) to bloom (Meijón et al., 2010b unpublished). Furthermore, for these cultivars results have shown (Meijón et al., 2010b unpublished) that photoperiod is also the main regulating factor of floral induction.

In summary, given the different sensitivities of azalea cultivars (Bodson, 1983; Väiölä et al., 1999), defining the specific photoperiod and temperature requirements in the different cultivars of azalea not only improves flowering but also facilitates promotion of flowering in periods in which the azalea would otherwise naturally be in other phenological stages such as vegetative growth or rest.

APPLICATION OF GROWTH RETARDANT IN AZALEA

Chemical growth retardants are used primarily to control stem elongation of ornamental plants while they are growing in containers rather than after the plants are transplanted into the landscape. All these chemicals act by inhibiting gibberellin (GA) biosynthesis. Also, in a number of woody perennial species, a decrease in GA concentrations in the apical meristems is required for floral initiation to occur. In *Rhododendron* sp, applied gibberellins inhibit flowering and gibberellin biosynthesis inhibitors promote flowering (Sharp et al., 2010).

The effectiveness of the regulator depends on the chemical nature of the compound and the cultivar to which it is applied (Smit et al., 2005). This must be taken into consideration when generalizing about the application of a product to a particular species, in addition to the impact of other important variables such as the age of the plant, nutritional status, plant health, environmental conditions, method of application and timing of plant growth regulator application, all of which are known to play a decisive role in the results (Halmann, 1990; Basra, 2000).

To date, four different types of such inhibitors are known (Fig 1) (Rademacher, 2000):

1) Onium compounds, such as chlormequat chloride, mepiquat chloride, chlorphonium and AMO-1618, which block the cyclases copalyl-diphosphate synthase (CDP) and ent-kaurene synthase involved in the early steps of GA metabolism
2) Compounds with an N-containing heterocycle, e.g. ancymidol, flurprimidol, tetcyclacis, paclobutrazol, unicazole-P, and inabenfide. These retardants block cytochrome P450-dependent monooxygenases, thereby inhibiting oxidation of *ent*-kaurene into *ent*-kaurenoic acid.
3) Structural mimics of 2-oxoglutaric acid, which is the co-substrate of dioxygenases, which catalyze the late steps of GA formation. Acylcyclohexanediones, e.g. prohexadione-Ca and trinexapac-ethyl and daminozide, particularly block 3β-hydroxylation, thereby inhibiting the formation of highly active GAs from inactive precursors.
4) 16,17-Dihydro-GA$_5$ and related structures, which most likely act by mimicking the GA precursor substrate of the same dioxygenases.

In *Rhododendron* sp, the most often used class of chemicals are triazoles such as paclobutrazol and unicazol. Daminozide and chlormequat, in some experiments, had good effects on flower bud formation, but the effect on shoot elongation, especially with chlormequat, was sometimes insufficient and multiple sprays were required (Marosz and Matysiak, 2005). Both chemicals also proved to be highly phytotoxic treatments at high dose (Meijón et al., 2009). The effects of trinexapac-ethyl and prohexidione calcium on flower bud initiation and internode elongation on azalea plants are not yet known.

In a study investigating the effectiveness of different retardants (paclobutrazol, daminozide and chlormequat) in Blaauw's Pink and Johanna cultivars of azalea which were selected as contrasts, Meijón et al., 2009, found that plants treated with paclobutrazol, or daminozide, became more attractive through increased intensity of green leaves and rounding of the plant shape (Figure 2). Growth regulators induce an increase in the content of chlorophyll in the leaves giving plants more resistance to winter conditions, in addition to inducing the formation of large, rounded leaves of a more intense green (Thakur et al., 2006).

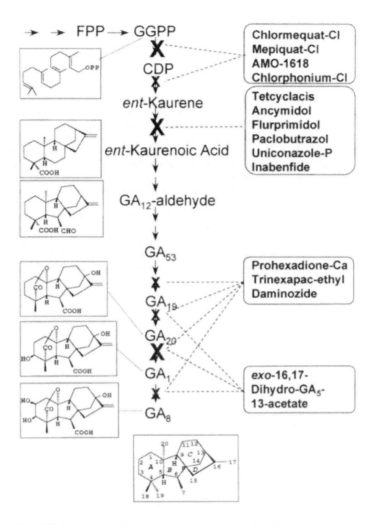

Figure 1. Simplified scheme of biochemical steps involved in GA biosynthesis and points of inhibition by plant growth retardants (X, x = major and minor activity, respectively). (Taken from Rademacher, 2000).

Chlormequat chloride soil drench application whilst effective for controlling plant growth and flowering in Blaauw's Pink and Johanna has proved highly phytotoxic at the doses at which it is effective for growth control, producing a mortality rate of approximately 35%. Low doses in soil drench application and administration by foliar spray when also tested did not have a significant effect on these cultivars (Meijón et al., 2009). Conversely,

Cathey (1975) describes chlormequat chloride as a broad spectrum growth regulator and very effective in soil drench application suggesting the need to determine the optimal dose of this product in drench application.

Figure 2. Image of the two cultivars of azaleas treated with different chemicals. (a) Control Blaauw's Pink, (b) Daminozide Blaauw's Pink, (c) Chlormequat Chloride Blaauw's Pink, (d) Paclobutrazol Blaauw's Pink, (e) Control Johanna, (f) Daminozide Johanna, (g) Chlormequat Chloride Johanna, (h) Paclobutrazol Johanna. (Taken from Meijón et al., 2009).

Although in Blaauw's Pink, daminozide was found to produce the best floral density and longer duration of flowering, at the applied dose, it caused torquing of petals and sepals, in addition to merging the stamens, thus rendering plants non-commercial (Meijón et al., 2009). This floral deformation persisted over three successive crops following treatment, an effect which has also been observed in other woody plants (Gent, 1997). Furthermore, Gent found that daminozide did not appear to be metabolized by some woody species. The decrease in the concentration of growth regulator in plant tissue appeared to be primarily the result of dilution by new biomass. In herbaceous plants, where the biomass doubles every week or two, this process may be rapid, but growth of woody plants is generally slower and often occurs in discrete flushes rather than continuously and thus the dilution of growth regulators by new biomass would take longer. Permanence of growth retardation is a characteristic of most plant growth retardants though it is especially pertinent in the case of triazoles where the reduction in stem elongation resulting from their use can last for several years in woody plants (Gent, 1995, 1997, 2004).

Paclobutrazol, in addition to providing optimum results as regards plant shape and leaf quality in Blaauw's Pink and Johanna, has also been found to be the most effective growth inhibitor in terms of improving flowering and in addition did not lead to morphological alterations, all of which strongly recommends its use in commercial production of these cultivars to improve quality and minimize costs (Meijón et al., 2009).

METHODS OF EVALUATION OF QUALITY IN ORNAMENTAL PLANTS

As a result of the increasing competition in the flowering potted plant market, there is an increasing demand for higher quality azaleas. Post-production tests are necessary to study the effect of different growth treatments on the quality of vegetative growth and flowering.

Lootens et al., (2000) developed a method for rapidly and objectively evaluating flowering. This was achieved by taking images from the top of the plants twice a week. Each time the image of a plant was taken, the area of the whole crown as well as the area of the flowers was measured. The ratio of the flower area divided by the crown area (i.e. the color patch) gives a measure of the flowering stage. A flowering curve results, from which important flowering parameters can be deduced: (1) the forcing time (time needed to force the plant into a saleable product, i.e. number of days until 30 % color patch is reached), (2) the flowering period (number of flowering days, i.e. number of days during which the plants has a color patch of more than 60%), and (3) the maximum color patch (i.e. maximum area of leaves covered by flowers, indicating the uniformity of flowering of plant). With standardized forcing conditions, objective quantification of the effects of different cultivation treatments on flowering quality is possible.

The correct architecture for an ornamental plant involves many factors, many of which are difficult to evaluate using classic biometry. It is crucial to find a fast and objective method for the evaluation of plant shape, and for comparing the effectiveness of various treatments.

Although quality in ornamental plants is a difficult aspect to define, the term is commonly used in the marketing of this type of plant. Sometimes the H/D parameter is employed as a development index in azalea (Bird and Conner, 1999). If this ratio takes values less than one, the plant will have a rounded shape and therefore will reach higher commercial quality. But when it takes values close to, or above, unity, it indicates that the plants have grown with too much apical dominance. Nevertheless, this index sometimes provides erroneous information since the visual appearance of some plants with values above unity is far from the ideal as they take on an irregular shape which is not commercially desirable (Meijón et al., 2009). To summarize, the use of traditional biometry parameters of the H/D ratio type leads to conclusions that do not always correspond with reality. Other parameters used, such as shoot growth (Meijón et al., 2009), are more closely related to plant architecture; however, this parameter is not a quality indicator, as it only really indicates a growth trend.

A protocol of image analysis has been developed that has shown itself to be a very efficient tool in quantifying and evaluating images that are very complicated to study (Meijón et al., 2009). In the protocol developed by Meijón et al. (2009), roundness, roughness and leaf coverage parameters were those selected to evaluate plant architecture:

- *Roundness is* the ratio between the real plant surface and the approximation to circularity of the plant surface. This ratio presents values between 0 and 1, 1 being a perfect circle ratio.

$$Roundness = \frac{Acircle}{Aobject*1.064} = \frac{\eta \dfrac{P^2}{4\eta^2}}{A*1.064} = \frac{P^2}{4\eta*A*1.064}$$

A = Area (cm^2); P= Perimeter (cm); η = pi; 1.064= Imaging Correction Factor

- *Roughness* is the relationship between the plant perimeter and the plant profile, providing information on the degree of branching which is beyond a length considered to give a homogenous profile. This ratio presents values equal to or greater than 1, 1 being the reference number.

$$Roughness = \frac{Pt}{Pcvxo}$$

Pt = Total perimeter (cm); Pcvxo = Convex perimeter (cm)

- *Leaf coverage* is the direct measurement, as a percentage, of the plant area covered with branches and leaves.

Once parameters which are easily acquired and interpreted are defined by image analysis, the efficacy of the products tested to control azalea development can be readily established.

CONCLUSION

The application of plant growth retardants, especially paclobutrazol, is the most effective strategy for controlling architecture and flowering in azalea. Methodologies based on the manipulation of photoperiod and temperature, as well as strategies based on nutritional or water pressure deficits require specific protocols for each species and even for each cultivar which increases production costs. However, the seasonality of bloom can only be achieved by modifying the photoperiod and temperature adapted to the requirement of each species in each development stage.

Through image analysis it is possible to objectively evaluate the commercial quality of azalea plants, in terms of flowering and vegetative development, in addition to deciding which treatment is the more effective growth retardant.

REFERENCES

Basra AS, 2000. *Plant Growth Regulators in Agriculture and Horticulture. Their Role and Commercial Uses.* The Haworth Press Inc., Binghamton, NY.

Bird RE, Conner, J.L., 1999. Container grown azalea response to sumagic sprays. In: *Proceedings of SNA Research Conference,* vol. 44. pp. 274–276.

Bodson M, 1983. Effect of photoperiod and irradiance on floral development of young plants of a semi-early and a late cultivar azalea. *Journal of American Society of Horticulturae Science*, 108(3): 382-386.

Cameron RWF, Harrison-Murray RS, Scott MA, 1999. The use of controlled water stress to manipulate growth of container-grown Rhododendron cv Hoppy. *J. Hort. Sci. Biotech.* 74:161-169.

Cathey HM, 1975. Comparative plant growth-retarding activities of ancymidol with ACPC, phosfon, chlormequat, and SADH on ornamental plant species. *HortScience* 1 (3), 204–216.

Corbesier, L., Prinsen, E., Jacqmard, A., Lejeune, P., Van Onckelen, H., Périlleux, C. and Bernier, G. (2003) Cytokinin levels in leaves, leaf exudate and shoot apical meristem of Arabidopsis thaliana during floral transition. *Journal of Experimental Botany*, 54, 2511-1517.

Edmondson RN, Parsons NR, 2002. Artificial neural networks for pot-plant quality. Final Report for DEFRA project HH1529SPC, DEFRA.

Edmondson RN, Parsons NR, 2005. The measurement and improvement of robust bedding plant quality and the use of digitial imaging for quality assessment. HDC Project Report PC200, HDC.

Fernández MD, Hueso JJ, Cuevas J, 2010. Water stress integral for successful modification of flowering dates in "Algerie" *loquat. Irrig Sci* (2010) 28:127–134.

Fletcher, R.A., Gilley, A., Sankhla, N., Davis, T.D., 2000. Triazoles as plant growth egulators and stress protectants. *Hortic. Rev.* 24, 55–138.

Gent MPN, 1995. Paclobutrazol or uniconazol applied early in the previous season promotes flowering of field-grown Rhododendron and Kalmia. *J. Plant Regul.* 14, 205–210.

Gent MPN, 1997. Persistence of triazole growth retardants on stem elongation of Rhododendron and Kalmia. *J. Plant Regul.* 16, 197–203.

Gent MPN, 2004. Efficacy and persistence of paclobutrazol applied to rooted cuttings of Rhododendron before transplant. *HortScience* 39 (1), 105–109.

Grindal G, Junttila O, Reid JB, Moe R, 1998. The response to gibberellins in Pisum sativum grown under alternating day and night temperature. *Journal Plant Growth Regulation,* 17:161-167.

Halmann M, 1990. Synthetic plant growth regulators. *Adv. Agron.* 43, 47–105.

Hansen CW, Lynch J (1998). Response to phosphorus availability during vegetative and reproductive growth of chrysanthemum: II Biomass and phosphorus dynamics. *Journal of the American Society for Horticulturae Science.* 123(2): 223-229.

Islam N, Patil GG, Gislerød HR, 2005. Effect of photoperiod and light integral on flowering and growth of Eustoma grandiflorum (Raf.) *Shinn. Sci. Hortic.* 103, 441–451.

Jackson S.D., 2009. Plant responses to photoperiod. *New Phytologist,* 181: 517-531.

Lootens, P., Vandecastelle, P., Heursel, J., 2000. Evaluation of flowering quality of Rhododendron simsii cultivars through image analysis. Acta Hortic. 517, 329–333.

Marosz A, Matysiak B, 2005. Influence of growth retardants on growth and flower bud formation in Rhododendron and azalea. *Dendrobiology* 54, 35–40.

Meijón M, Cañal MJ, Rodríguez R, Feito I, 2010a. Epigenetic and physiological effects of plant growth regulators and chemical pruners on the floral transition of azalea. Under review in Planta.

Meijón M, Feito I, Rodríguez R, Cañal MJ, 2010b. Promotion of flowering and epigenetic monitoring of floral induction and bloom in azalea. Under review.

Meijón M, Rodríguez R, Cañal MJ, Feito I, 2009. Improvement of compactness and floral quality in azalea by means of application of plant growth regulators. *Sci. Hortic.* 119, 169-176.

Moe R, Heins R, 1990. Control of plant morphogenesis and flowering by light quality and temperature, Acta Horticulturae 272.

Navarro A, Sánchez-Blanco MJ, Bañon S, 2007. Influence of paclobutrazol on water consumption and plant performance of *Arbutus unedo* seedlings. *Sci. Hortic.* 111, 133–139.

Rademacher W, 2000. Growth retardants: effects on gibberellin biosynthesis and other metabolic pathways. *Annual Review Plant Physiology and Plant Molecular Biology*, 51, 501-531.

Richards D, Wilkinson RI, 1984. Effect of manual pinching, potting-on and cytokinins of branching and flowering of Camellia, Rhododendron and Rosa. Sci. Hortic. 23, 75–83.

Ristvey AG, Lea-Cox LD, Ross DS, 2007. *Journal of the American Society for Horticultural* Sciences. 132(4): 563-571.

Sharma GC, 1979. Controlled-release fertilizers and horticultural applications. *Sci. Hortic.* 11(2): 107-129.

Sharp RG, Else MA, Cameron RW, Davis WJ. 2009. Water deficits promote flowering in Rhododendron via regulation of pre and post initiation development. *Scientia Horticulturae*, 120:511-517.

Sharp RG, Else MA, Davis, WJ Cameron, 2010. Gibberellin-mediated suppression of floral initiation in the long-day plant Rhododendron cv. Hatsugiri. Scientia Horticulturae, in press.

Shu, LJ, Sanderson KC, Williams JC, 1981. Comparison of several chemical pinching agents on greenhouse forcing azaleas, Rhododendron cv. *J. Am. Soc. Hortic. Sci.* 106 (5), 557–561.

Skinner PW, Matthews MA, 1989. Reproductive development in grape (Vitis vinifera L.) under phosphorus-limited conditions. *Sci. Hortic.* 39, 49-60.

Smit M, Meintjes JJ, Jacobs G, Stassen PJC, Theron KI, 2005. Shoot growth control of pear trees (Pyrus communis L.) with prohexadione-calcium. *Sci. Hortic.* 106, 515–529.

Stern RA, Naor A, Bar N, Gazit S, Bravdo B-A, 2003. Xylem-sap zeatin-riboside and dihydrozeatin-riboside levels in relation to plant and soil water status and flowering in "Mauritius" lychee. *Scientia Horticulturae*, 98:285-291.

Thakur, R., Sood, A., Nagar, P.K., Pandey, S., Sobti, R.C., Ahuja, P.S., 2006. Regulation of growth of Lilium plantlets in liquid medium by application of paclobutrazol or ancymidol, for its amenability in a bioreactor system: growth parameters. *Plant Cell Rep.* 25, 382–391.

Väinöla A, Junttila O, Rita H, 1999. Cold hardiness of rhododendron cultivars grown in different photoperiods and temperatura. *Physiologia Plantarum,* 107: 46-53.

Wilkie JD, Sedgley M, Olesen T, 2008. Regulation of floral initiation in horticultural trees. *Journal of Experimental Botany*, 59(12): 3215-3228.

In: Flowering Plants
Editor: Jeremy J. Tellstone

ISBN: 978-1-61324-653-5
© 2011 Nova Science Publishers, Inc.

Chapter 7

THE EVOLUTION OF CARNIVORY IN FLOWERING PLANTS[*]

Chris Thorogood
School of Biological Sciences, University of Bristol, UK

ABSTRACT

Carnivorous plants are extremely derived and many species have evolved distinct modifications in gross vegetative structures to complement their derived mode of nutrition. This divergence in morphological features has deprived taxonomic scientists of traits with which to delineate the evolutionary relationships of carnivorous plant families with their non-carnivorous ancestors. Carnivorous plants have long aroused the interest of botanists, and were used as model systems for physiological studies of movement and secretion in plants, pioneered by Darwin (1875). Studies on the evolution of carnivorous plants were depauperate in the literature for most of the 20th Century, and their origins remained largely speculative and unsubstantiated. However, advances in molecular technology and the advent of gene sequencing and molecular phylogenetics towards the end of the 20th Century have vastly improved our understanding of the evolution of flowering plants, including the origins of carnivory.

---

[*]A version of this chapter was also published in The Malaysian Nepenthes: Evolutionary and Taxonomic Perspectives, by Chris Thorogood, published by Nova Science Publishers, Inc. It was submitted for appropriate modifications in an effort to encourage wider dissemination of research.

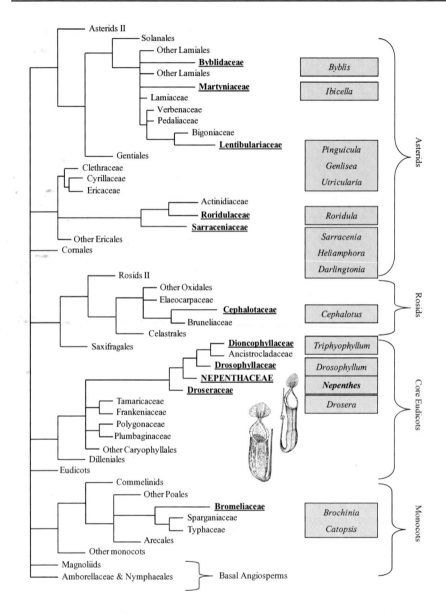

Figure 1. Positions of the carnivorous plant families in the angiosperm phylogeny adapted from Ellison and Gotelli (2009), Stevens (2007) and Müller et al. (2006). Note the branch lengths are not meaningful.

Introduction

Carnivory is now established to have evolved independently at least six times in five angiosperm orders (Figure 7) (Ellison and Gotellii, 2009), contrary to speculation that all carnivores arose from a common ancestor (Croizat, 1960; Heads, 1984). At present, approximately 600 species of carnivorous plant are recognised, from 11 families and 19 genera, and include representatives from both the core eudicots and the monocots (Heubl et al., 2006). Carnivorous plants show remarkable morphological and physiological convergence in the face of their independent origins, which has obscured their affinities in then past. Darwin (1859) first recognised the importance of homology in the evolution of flowering plants, and the convergent evolution of traits across unrelated taxa. The view of multiple independent origins of carnivory has been widely accepted, however speculation by Croizat (1960) suggested that all carnivorous plants arose from a common ancestor, and that they comprised a relatively basal group in the angiosperm lineage. This hypothesis has been consistently refuted by results from analyses based on molecular and phylogenetic techniques. Many of Darwin's observations laid the foundations for research into carnivorous plants, and some of his predictions on homology were borne out by modern analyses. Other recent results are more surprising, for example Darwin's assertion that *Nepenthes* are 'not at all related to the Droseraceae' which has been refuted by such recent phylogenetic analyses. This chapter examines research into the evolutionary origins of *Nepenthes* with a focus on phylogenetic studies.

The Advent of Phylogenetics

It is established that common ecological factors have led to extraordinary degrees of convergence among carnivorous plants, which obscured important evolutionary differences in the past (Ellison and Gotelli, 2001). For example, morphological adaptations to the carnivorous syndrome were the basis for classifying the families Nepenthaceae, Sarraceniaceae and Droseraceae in the artificial order Sarraceniales. Thorne (1992) in his classification of the flowering plants prior to the advent of large-scale phylogenetic revision, placed the Nepenthaceae and Sarraceniaceae near the Theales, and dissociated them the Droseraceae which was placed in the Rosales (suborder Saxifragales). The first substantiated classification of the carnivorous plants, that revealed the independence of their evolutionarily origins, and of their distinct syndromes of trapping mechanisms employed cladistic analyses based on *rbc*L gene sequence data (Albert et al., 1992). The authors selected 100 eudicotyledonous taxa plus additional carnivorous species from 72 families, from which cladistic analyses indicated as many as seven independent origins of the carnivorous syndrome. This study provided a phylogenetic framework from which more robust inferences about the evolution of carnivory could later be made. Contrary to Darwin's predictions, this study indicated the polyphyly of structurally similar flypaper traps, and also identified their potential importance as the ancestral state from which other trap forms evolved multiple times. In fact, two of the three distantly related lineages of pitcher plant were identified to be independently closely related to flypaper trap families. Importantly the genus *Nepenthes* was revealed to be associated with the Droseraceae – an unpredicted result which was exemplary of structural evolution in a monophyletic lineage.

RESOLVING RELATIONSHIPS WITHIN THE CARYOPHYLLALES

The evolutionary shift to a carnivorous life history in flowering plants was associated with extreme alterations in gross morphology that was a breeding ground for speculation and taxonomic confusion. Similar problems arose in the classification of parasitic plants, for example the world's largest flower *Rafflesia* (Rafflesiaceae), the phylogenetic affinities for which have only recently been resolved, and which surprisingly nest in the Euphorbiaceae – a family characterised by its minute flowers (Davis et al., 2007). In traditional taxonomic treatments of the flowering plants, the families Droseraceae and Nepenthaceae were grouped in several eudicot orders, which variously indicated relationships with the Euphorbiales, Caryophyllales Saxifragales, Theales and the Violales (Cronquist, 1988, Thorne, 1992). Many authors placed the Droseraceae and Nepenthaceae in association with the Sarraceniaceae – a family of North American pitcher plants (Engler 1898; Cronquist, 1988). Takhtajan (1969) removed the Sarraceniaceae from the Nepenthaceae, and Thorne (1992) also separated these families from the Droseraceae. Clearly these carnivorous families suffered a confused taxonomic history, and morphological features of ambiguous ancestry such as gross floral traits, coupled with a peculiar evolutionary convergence of

The Evolution of Carnivory in Flowering Plants 157

features associated with carnivory seem to have confused systematists (Williams et al., 1994). Cladistic analyses based on *rbc*L gene sequence data later included the Droseraceae, Nepenthaceae and Sarraceniaceae. These analyses identified the independent origins of carnivory among the angiosperms, and importantly, grouped the families Nepenthaceae and Droseraceae (Albert et al., 1992) with the Caryophyllids, and separated these families from the Sarraceniaceae, which was allied with the Ericales. This early study provided a phylogenetic framework to which more taxa could be added, and from which the relationships between carnivorous taxa and their non-carnivorous ancestors could be better resolved. Williams et al. (1994) investigated the molecular support for the monophyly of the Droseraceae, and the phylogenetic relationships of this family within the eudicots, also using *rbc*L gene sequence data. This analysis, which used 100 species including families of subclasses Rosidae, Hamamelidae, Dillenidae and Caryophyllidae, placed the Droseraceae in the same clade as the Caryophyllidae and the Nepenthaceae. This study corroborated previous evidence that the Nepenthaceae and the Droseraceae were related, and also included morphological and phytochemical data for greater support, but failed to resolve the exact relationship between these families. The position of *Drosophyllum* (Drosophyllaceae) – a monotypic genus of carnivorous plants represented by *D. lusitanicum* native to the western Mediterranean Basin, with an unusual habitat preference for dry, alkaline substrates, remained unclear.

Fay et al. (1997) included additional taxa in an effort to resolve relationships within the expanded Caryophyllids, for example sequence data from the Physenaceae, Asteropeiaceae, Frankeniaceae, Tamaricaceae, and, importantly from the Ancistrocladaceae and Dioncophyllaceae. This investigation suggested that *Drosophyllum* was in fact clearly phylogenetically separated from the Droseraceae – a relationship that was previously unclear. In addition, *Triphyophyllum* (Dioncophyllaceae) appeared to be closely related to the Ancistrocladaceae – a non-carnivorous family of old-world lianas belonging to the single genus *Ancistrocladus*. Finally Lledó et al. (1998) carried out a systematic study of the Plumbaginaceae based on *rbc*L sequence data which included 42 taxa, and revealed for the first time a clade including all carnivorous taxa, comprising: Droseraceae, Dioncophyllaceae, Nepenthaceae, and Ancistrocladaceae, in which *Nepenthes* was basal. In addition, physiological characteristics such as the production of plumbagin which is characteristic of the Droseraceae, Nepenthaceae and Dioncophyllaceae (Harborne 1967) appeared to lend support to the phylogenetic placement of these families.

Although early investigations into the Caryophyllales made great progress in elucidating the molecular systematic relationships of this group, most studies were based on single, relatively conserved markers, particularly *rbc*L gene sequences. While such markers provide support for higher taxonomic ranks, they often lack the desired resolution or support for more closely related taxa (Bailey et al., 2004). In addition, conflicting topologies arose from the exploitation of alternative markers (Meimberg et al., 2000), for example the placement of the Nepenthaceae, that was sister to the Caryophyllales according to the plastid gene sequence *ORF2280* (Downie et al., 1997), which was at odds with *18S* rDNA sequence data, which suggested a more distant relationship (Soltis et al., 1997). The relationships between the Droseraceae, Dioncophyllaceae, Nepenthaceae, and Ancistrocladaceae, and their affinities with the Caryophyllales, Plumbaginaceae and Polygonaceae were weakly resolved, therefore to overcome the problem of contradictory phylogenetic reconstructions, Meimberg et al. (2000) produced an independent data set based on *matK* sequences, which have a much higher mutation rate than traditionally used markers such as *rbc*L. Contrary to previous *rbc*L-based analyses, the *matK* tree topology suggested that the Nepenthaceae was a derived family, and sister to the families Ancistrocladaceae and Dioncophyllaceae. This analysis, which included all carnivorous families, indicated that carnivory was indeed of monophyletic origin, and that the syndrome was partially lost in the Dioncophyllaceae, and completely lost in the Ancistrocladaceae. The analysis also refuted the equivocal basal position of the Nepenthaceae in the carnivorous clade in the *rbc*L tree from the systematic study of the Plumbaginaceae (Lledó et al., 1998). The *matK*-based phylogeny also supported the separation of *Drosophyllum* from the Droseraceae, as well as the close relationship between the Ancistrocladaceae and Dioncophyllaceae in agreement with previous studies (Fay et al., 1997; Lledo et al., 1998). Importantly, the study by Meimberg et al. (2000) presented the first unambiguous case for the monophyletic origin of the carnivorous syndrome in the Caryophyllales. Molecular phylogenetic analyses of the Caryophyllales based on nuclear 18S rDNA and plastid *rbc*L, *atpB* and *matK* DNA sequences all supported the monophyletic origin of carnivory, and with increased support for deeper branches of the phylogeny, and divided most taxa in the order into two main clades – the 'core' and 'noncore' Caryophyllales, as recognised by the Angiosperm Phylogeny Group (APG, 1998); the former group comprising those families recognised previously, and the latter comprising the additional families included upon redefinition of the order, including carnivorous taxa (Cuénoud et al., 2002). While the core group corresponded with previous circumscriptions of the order, the delineation of noncore group comprising Polygonaceae, Plumbaginaceae, Frankeniaceae and Tamaricaeae as well as the carnivorous families was the result of advances in molecular techniques. The monophyly of the noncore Caryophyllales was well resolved by the combined analysis, in contrast with previous analyses based on *rbc*L (Savolainen et al., 2000). Diagnostic features of the noncore caryophyllid clade include: secretory cells producing plumbagin, stalked glandular trichomes, basal placentation and starchy endosperm (Judd et al., 1999). Taken together, the combined gene sequence-based phylogenetic analyses and morphological data provide strong evidence for the monophyly of the carnivorous clade in the Caryophyllales, which has provided a platform for hypotheses about the evolution of carnivory in *Nepenthes* and related taxa.

THE EVOLUTIONARY SHIFT TO CARNIVORY IN THE CARYOPHYLLALES

Early studies based on single gene sequence, morphological and phytochemical data lacked the statistical power necessary to unravel taxonomic relationships to a fine resolution, but provided systematists with the architecture for solving the mystery of how carnivorous plants evolved from their non-carnivorous ancestors. Over 95 % of the ~600 species of carnivorous plant in the plant kingdom are currently placed in the Caryophyllales or in the Lamiales (Ellison and Gotelli, 2009). The Caryophyllales comprise a large and diverse order of core eudicots besides the carnivorous plants which include succulents, cacti and plants with a plethora of specialisations for saline, xeric and acidic waterlogged environments. Many of the environments in which representatives of the Caryophyllales most frequently occur require specialised mechanisms of nutrient uptake, which probably predisposed this order to carnivory. Indeed, the phyletic co-occurrence of the carnivorous habit together with an unusual secondary chemistry in the Ericales from which the carnivorous Sarraceniaceae evolved supports this hypothesis (Williams et al., 1994). Meimberg et al. (2000) produced phylogenetic analyses including all carnivorous families in the Caryophyllales which indicates that the ancestral state linking all carnivores is analogous to that of *Drosophyllum* and the modern Droseraceae. Therefore the Nepenthaceae along with its sister groups *Drosophyllum,* Ancistrocladaceae and Dioncophyllaceae appear to have evolved from an ancient Droseraceae-like lineage (Figure 8; Figure 9).

The gland is a salient feature of the carnivorous syndrome, and may have been pivotal in the evolution of carnivorous plants by virtue of the improved efficiency of nutrient absorption (Juniper et al., 1989). Therefore a possible starting point for the evolutionary transition to carnivory in the Caryophyllales could have been adhesive glands, which are common among species in the Plumbaginaceae. Insect fixation efficiency may have improved via the development of glands that secrete and also absorb, driving the selective advantage for the colonisation of nutrient-poor environments (Meimberg et al., 2000). Schlauer (1996) discussed the possibility of physiological traits of extant members of the Caryophyllales as precursors to carnivory in this group. Specialised lime and mucilage secreting glands on the leaves of members of the Plumbaginaceae show a superficial resemblance to sessile glands on the leaf surfaces of *Ancistrocladus* and *Nepenthes,* the genetic associations of which were confirmed by early phylogenetic analyses. Lledó et al. (1998) provided molecular support for the familial relationships in the caryophyllid clade, and discuss the life history strategies associated with the Plumbaginaceae that allow them to tolerate nitrogen-poor soils, saline conditions and drought. The family is highly stress-tolerant, and plants adapted to dry and saline areas are characterised by the accumulation of osmoprotectants and the presence of glandular mechanisms for salt secretion (Hanson et al., 1994). These salt glands are also present in other related families such as the Chenopodiaceae, Frankeniaceae and Tamaricaceae, and members of the Caryophyllales often show a succulent habit coupled with crassulacean acid metabolism and C_4 photosynthesis (Lledó et al., 1998). Schlauer (1996) suggested that the calyx glands of *Plumbago* (Plumbaginaceae) which share anatomical features with flypaper carnivores in the Caryophyllales (Rachmilevitz and Joel, 1976) may be similar to precursor prototypes of tentacles, and formulated an hypothesis for the evolution of carnivory. This hypothesis was characterised by (i) the evolution of bristles on flower sepals as an aid to seed dispersal; (ii) the emergence of glandular properties on bristles in relation to pollination; (iii) the utilisation of nutritive properties of insects inadvertently caught on glandular trichomes; (iv) the translocation of tentacles to leaf surfaces. Further evolution of particular trapping syndromes from this putative basal trait would then be driven by particular selective advantages associated with particular habitats, and or prey. Morphological traits support the molecular data indicating the origin of *Nepenthes* from an ancient Droseraceae-like lineage, such as the occurrence of pollen tetrads, small chromosomes, the presence of particular napthoquinones, filiform seeds with starch, anther morphology and variable stamen number (Heubl and Wistuba, 1995; Meimberg et al., 2000).

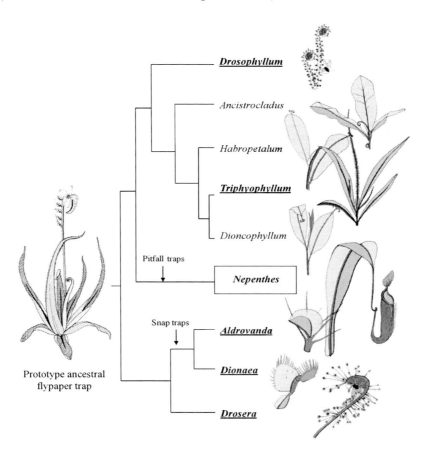

Figure 2. Phylogenetic relationships of the monophyletic carnivorous clade within the non-core Caryophyllales inferred from parsimony analyses of partial *mat*K sequences (Meimberg et al., 2000) and combined *rbc*L, *atp*B, *mat*K, and 18S rDNA (Cameron et al., 2002), adapted from Heubl et al. (2006). A hypothetical diagrammatic representation of the putative flypaper precursor to this clade is provide; carnivorous taxa are depicted in bold and underlined.

Mapping traits onto the current most reliable four-gene phylogeny supports the hypothesis of a common ancestor with adhesive organs with multicellular glands and suggest the precursor lineage had adaxially circinate leaves and a weak root system, as well as a habitat preference for nutrient-poor terrestrial environments (Figure 8) (Heubl et al., 2006). During the evolution of the complex trapping mechanism of *Nepenthes,* a leaf transformation in which the multicellular glands of the putative Droseraceae-like progenitor became embedded in the pitcher, and partially covered by epidermal tissue may have occurred (Meimberg et al., 2000; Heubl et al., 2006). Heubl et al. (2006) suggest that the epiascidiation (folding of glandular leaves followed by marginal fusion) could have conferred an immediate selective advantage by virtue of a novel capacity for water storage and nutrient absorption in mineral-deprived environments. The short branch lengths of the *matK* sequence-based phylogenetic reconstructions suggest that the Nepenthaceae and its sister group the Ancistrocladaceae underwent a rapid diversification driving their evolutionary history (Heubl et al., 2006).

Figure 3. Members of the monophyletic carnivorous clade (including non-carnivorous representatives) within the non-core Caryophyllales; A. *Drosophyllum;* B. *Habropetalum*; C. *Ancistrocladus*; D. *Dioncophyllum;* E. *Triphyophyllum;* F. *Nepenthes.*

The Evolution of Carnivory in Flowering Plants 163

Hypotheses about the origins of carnivory cannot, of course be tested empirically, but comparative investigations into the evolutionary development ('evo-devo') of genes in carnivorous plants and their putative non-carnivorous ancestors may substantiate hypotheses related to the evolution of the carnivorous syndrome in flowering plants in the future. Evo-devo was granted its own division in the Society for Integrative and Comparative Biology in 1999, and was initiated as a common ground for evolutionary and developmental biologists (Goodman and Coughlin, 2000). The basis for the discipline is that evolutionary biologists seek to understand how organisms evolve, and the roots of such evolution are found in the developmental mechanisms that control shape and form. *Nepenthes* offers an attractive model for evo-devo studies because of the clear mismatch between functional and developmental modularity, i.e. different parts of a single ancestral developmental module (the leaf) have distinct functions – photosynthesis (the leaf blade) and carnivory (the pitcher) (Breuker et al., 2006). Studying the evolutionary and developmental biology of *Nepenthes* pitchers may therefore open up a promising avenue of research into understanding the evolutionary origins of carnivory in the plant kingdom.

In understanding the origins of carnivory, it is important to establish the adaptive value of the syndrome, and researchers have focussed on the nutritional benefits (Adamec 1997; Ellison and Gotelli, 2001). To elucidate the evolutionary pathway to carnivory in the caryophyllids, scientists must consider other plant families in which the syndrome has evolved. Families in which carnivory is considered to be a relatively recently evolved trait, or that also contain non-carnivores or 'protocarnivores' may also be useful models in deciphering the origins of carnivory. For example, the bromeliad *Brocchinia reducta* in the order Poales is generally agreed to be carnivorous on the basis of its unusual nectar-like fragrance attractive to insects and trichomes capable of absorbing amino acids from a pool of liquid formed by a cylinder of leaves in which insects drown (Givnish et al., 1984). This species shows a basic level of specialisation to the carnivorous syndrome, and lacks digestive glands; uniquely furthermore, the genus contains other non-carnivorous species, which offers the opportunity to research the origins of carnivory. Givnish et al. (1984) propose that the invasion of sterile, sunny savannas by the ancestor of *B. reducta* may have favoured the evolution of vertically inclined, brightly coloured leaves, preadaptive to carnivory. The adaptive shift to carnivory may have been accelerated by the secretion of sweet-smelling volatile compounds which are also produced by non-carnivorous *Brocchinia* species. The authors propose a cost-benefit model that predicts that carnivory is adaptive only in sunny, moist, nutrient-deprived environments. Looking towards the basal angiosperms and non-flowering plants may also be fruitful in the path to understanding the origins of carnivory in the plant kingdom; the liverworts which form cavities in the lobes of their leaves are promising candidates for investigating the carnivorous syndrome in the bryophytes (Heubl et al., 2006).

THE MISSING LINK TAXA

Albert et al. (1992) suggested that a key taxon for integrating the newly recovered Nepenthaceae - Droseraceae cluster was *Triphyophyllum* (Dioncophyllaceae) (Figure 9 E), a monotypic genus native to tropical western to Equatorial Africa, unavailable for sequencing at the time. This cryptic species presented a combination of *Nepenthes*-specific and *Drosera*-specific morphological traits, which became parsimonious once these groups were brought into phylogenetic proximity. *Triphyophyllum* is heterophyllous, and produces three distinct leaf types during different stages of maturation. The first leaves are laminate and without glands, and are followed by filiform leaves with both sessile and stalked glands, the latter secreting a sticky, acidic mucilage which has insect-trapping ability (Green et al., 1979; Marburger, 1979). A phase shift occurs during the maturation of the plant, in which juvenile rosettes elongate to form a liana up to 50 m high, with leaves that lack glands, but have paired terminal extensions of the midrib, similar to the tendrils of *Nepenthes*. The insectivorous habit may be significant in making possible the transition from the juvenile non-climbing shoots to the rapidly climbing mature plant (Green et al., 1979). *Triphyophyllum* shares vascularised sessile and stalked glands which secrete insect-trapping mucilage with the Droseraceae, and a distinct phase shift in life history from rosettes to lianas, terminal extensions of the midrib, and cortical vascular bundle anatomy with the genus *Nepenthes* (Marburger, 1979, Albert et al. 1992).

Triphyophyllum belongs to the Dioncophyllaceae which was described by Airy Shaw (1951) in his account of the family as 'one of the strangest groups of plants to be found in the vegetable kingdom'. *Dioncophyllum tholloni*, of the same family was first described in French Equatorial Africa, and was suggested to be affiliated with the Bixaceae and Passifloraceae by its founder (Baillon, 1890). Warburg (1893) then placed this unusual plant in the Flacourtiaceae (=Salicaceae), but remarked that the terminal extensions of the midrib, or 'tendrils' rendered it possible that the genus belonged to the Passifloraceae. Sprague (1916) reinvestigated the elusive plant on receiving

The Evolution of Carnivory in Flowering Plants 165

sterile material at Kew, and similarly referred the genus to the Passiflorales, but underscored the similarities between the paired terminal extensions of the midrib with the tendrils of *Nepenthes*. Airy Shaw (1951) explicitly hypothesised that *Dioncophyllum* was not at all related to the Flacourtiaceae, but showed similarities with both *Nepenthes*, and interestingly with the Droseraceae. He placed the Dioncophyllaceae within the Sarraceniales (now disintegrated), near the Nepenthaceae, and subdivided the family into three genera: *Habropetalum*, *Dioncophyllum* and *Triphyophyllum,* the latter species being carnivorous, and the former two species entirely non-carnivorous. Marburger (1979) examined the glandular leaf structure of *Triphyophyllum peltatum* and commented on the similarity between the glandular anatomy of this species with *Drosophyllum lusitanicum* (Drosophyllaceae, ex Droseraceae). Due to the suspected divergent origins of these species, the author proposed that the morphological and anatomical similarities were the product of convergent evolution, rather than evidence of an alliance between these species.

The increased resolution of ancestry of carnivorous taxa in the Caryophyllales based on *matk* sequence data provides a template with which to evaluate the assumption of Albert et al. (1992) that *Triphyophyllum* may be a link between carnivores in the Droseraceae-Nepenthaceae group (Meimberg et al., 2000). The *matk*-based phylogeny produced by Meimberg et al. (2000) indicates that the ancestral state linking the caryophyllid carnivores was similar to the flypaper syndrome of *Drosophyllum* and *Drosera*. Therefore Nepenthaceae and its sister groups Drosophyllaceae (Figure 9 A), Ancistrocladaceae (Figure 9 C) and Dioncophyllaceae (Figure 9 D) are likely to have evolved from an ancient Droseraceae-like lineage. Carnivory was then presumably lost in Ancistrocladaceae and Dioncophyllaceae with the exception of *Triphyophyllum*. The derived phylogenetic placement of these taxa that have lost carnivory in the carnivorous caryophyllid clade is an interesting phenomenon, given the common presumption that carnivory is an extremely derived trait (Ellison and Gotelli, 2001). In contrast with Albert et al. (1992), Meimberg et al. (2000) suggest that *Drosophyllum* may be the previously 'missing link' taxon integrating the families Droseraceae + Nepenthaceae and Ancistrocladaceae + Dioncophyllaceae, based on their *matk*-based phylogeny. Taken together, the molecular data suggest that *Drosophyllum* is distinct from the Droseraceae, and belongs in its own family, Drosophyllaecae. Moreover, in corroboration with morphological data, the Ancistrocladaceae and Dioncophyllaceae, which overlap in their geographic range in West Africa, are clearly closely related phylogenetically. These

166   Chris Thorogood

families share features such as a woody liana-like life history, hooks or tendrils for a climbing habit, peltate hairs, and petioles with sclerenchyma bundles (Albert and Stevenson, 1996).

PROTOCARNIVOROUS PLANTS

The fossil record of carnivorous angiosperms is depauperate and fragmentary, and serves only to demonstrate that the carnivorous plants were widespread from the beginning of the Tertiary period (Juniper et al., 1989). To elucidate the pathway to carnivory in carnivorous plants such as *Nepenthes,* it may be beneficial to look to extant examples of plants that may be at the evolutionary cusp of the transition from non-carnivory to carnivory. Representative carnivorous plant families that have been studied by scientists and cultivated by horticulturists alike have become iconic of the group. However the exact definition of carnivory in plants is ambiguous, and there are grounds to consider expanding the current group to include so-called protocarnivorous plants. These are plants that do not comply with all the criteria met by species traditionally described as carnivorous, or for which the potential for carnivory has not yet been fully investigated. Darwin (1875) suspected carnivory in a range of plants that produce adhesive glands such as *Saxifraga, Primula, Pelargonium, Erica* and *Mirabilis*, and other authors have similarly nominated species with sticky hairs as potential candidates for investigation (Simons, 1981). Early studies into the anatomy and physiology of *Petunia, Martynia* and *Lychnis* spp. revealed the presence of mucilage and digestive enzymes in the secretions from the hairs of these species which also appeared to readily capture and kill small insects (Mameli, 1916; Mameli and Aschieri, 1920; Zambelli, 1929). Perhaps more surprisingly, seeds of species in the Brassicaceae that occur in nutrient-poor environments appear to have the ability to trap, digest and absorb insects, and release chemicals that attract mosquito larvae (Barber, 1978). Simons (1981) discusses the variety in water-holding structures in the plant kingdom that may be similar to the evolutionary precursors of pitcher plants, which are best characterised by the epiphytes such as Bromeliads, which live perched on tree branches and have evolved the means to store water as a result of erratic water supplies. Many Bromeliad species trap dead leaves, insect remains and other organic debris. A variety of plant species besides Bromeliads possess water-holding structures that may serve a carnivorous function, such as *Drynaria, Billbergia* and *Dischidia* that require further investigation. In addition, the leaf lobes of many liverworts trap

The Evolution of Carnivory in Flowering Plants 167

water which contains microscopic organisms, and may be carnivorous (Simons, 1981; Heubl et al., 2006). Those of the tropical genera *Colura* and *Pleurozia* have produced simple pitchers, which organisms can force their way into, but which become trapped inside by virtue of inward-pointing flaps (Simons, 1981). The extent of carnivory in liverworts has been little-explored and remains undetermined, but investigating lower plants may be fertile ground for deciphering the steps to carnivory, and may complement evo-devo studies in higher plants.

Some plants have been classified as carnivorous relatively recently, based on rigorous experimental investigation. For example the genus *Heliamphora* endemic to the Precambrian sandstone formations in Venezuela were previously disregarded as being carnivorous on the basis of their lacking specialisations typical of the syndrome (Juniper et al., 1989). Jaffe et al. (1992) redefined the genus as being carnivorous, after describing traits such as attraction of prey through visual and chemical signals, trapping and killing of prey, digestion though secreted enzymes, and the absorption of nutrients. The genus *Heliamphora* is now generally accepted as being truly carnivorous. The situation in other plants is less clear; many plants traps insects as a by-product of pollination or though defence with apparently no selective advantage to the plant (Juniper et al., 1989). Glandular plant species are taxonomically widespread, and have evolved the secretion of mucilage as a defence against herbivory. This mucilage hinders insect movement and also traps small arthropods where they die of exposure or starvation (Spomer, 1999). Members of the genus *Passiflora* have been reported to have evolved modifications predisposing plants to insect capture. For example *P. adenopoda* produces sharp trichomes that impale herbivorous caterpillars (Gilbert, 1971). The flowers of *P. foetida,* a species that inhabits nutrient-poor environments, produce reticulate bracts which protect the developing buds and fruits from predation. These bracts also possess minute glands which exude secretions that trap insects. Radhamani et al. (1995) demonstrated the presence of protease and phosphatase activity, and the potential for the absorption of amino acids, indicating the dual functions of defence and carnivory of bracts in *P. foetida*. Similarly, the potential for protocarnivory was demonstrated for *Geranium viscosissimum* and *Potentilla arduta* and a range of other sticky plants by Spomer (1999). Experiments showed that both species were able to digest trapped protein, and proteinase activity was detected by the application of ^{14}C-labeled protein to the glandular surface. The author even suggested that sticky crops could be engineered to express protocarnivorous activity to reduce fertilizer application.

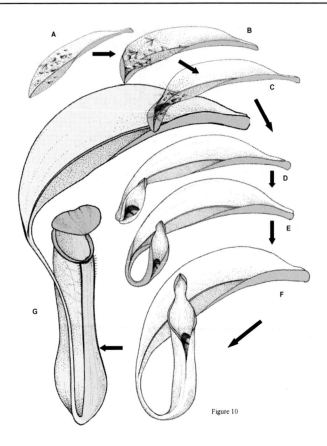

Figure 4. A hypothetical diagrammatic representation of the potential pathway (A– G) in the evolution of the *Nepenthes* pitcher from an ancient Droseraceae-lineage, involving apical foliar-folding followed by marginal fusion.

More recently, evidence of protocarnivory has been presented for triggerplants of the genus *Stylidium* (Styliadaceae). Darnowski et al. (2006) investigated the protocarnivorous tendencies of these plants which trap small insects with mucilage-secreting hairs on the peduncles and sepals of the inflorescences. Triggerplants are generally found in similar habitats to other carnivorous plant species belonging to the Droseraceae and Byblidaceae on permanently or seasonally wet nutrient-poor soils, predominantly in Australia. Specimens in herbaria were found to have trapped similar numbers of insects to conventionally accepted carnivorous plant species. In addition, protease activity was produced by glandular regions after the application of yeast extract. Other species that have complex symbiotic relationships with insects

The Evolution of Carnivory in Flowering Plants

may also be carnivorous to an extent, but require further investigation, for example the ant plants, in which insect food trapped by cohabiting ants provides nutrients for the plant (Simons, 1981), and *Roridula* – a South African genus of plants which benefit from the faeces of insects that feed on other insects trapped on the sticky leaves (Hartmeyer, 1998). Taken together, these studies suggest that protocarnivory may occur more widely in the plant kingdom than previously considered, but over a wider set of conditions than those indicated by the small group of well known carnivorous plant families (Spomer, 1999).

EVOLUTION OF THE PITCHER

The pitcher in the general sense is a taxonomically widespread phenomenon, and produced by ferns and lower plants as well as non-carnivorous flowering plants, for example the ant plants. Juniper et al. (1989) describe pitcher-like structures from various plant groups, for example those that are produced by plants as a mutation with no apparent function. Mutants of *Codiaeum variegatum* (Euphorbiaceae) apparently can produce pitcher-like tubular leaves, and the tips of a cultivar of *Ficus bengalensis* produce pocket-like structures. Similarly, the teasel *Dipsacus fullonum* has long been suspected of having carnivorous properties by virtue of the pools of water containing dead insects that occur at the leaf bases of this plant (Christy, 1923).

The production of pitcher-like apparatus via mutations similar to those observed in the species listed above, could confer an adaptive advantage to an ancestral flypaper lineage by improving trapping and absorptive capabilities. It has been suggested based on phylogenetic inference and the fossil record, that during the evolution of *Nepenthes*, leaf transformation caused the embedding and subsequent partial covering of glands by epidermal tissue in the putative Droseraceae-like progenitor (Meimberg et al., 2000; Heubl et al., 2006). Foliar-folding followed by marginal fusion may have conferred an immediate selective advantage following the novel capacity for water storage and nutrient absorption in mineral-deprived environments (Figure 10). The homology of trapping apparatus obscured the evolution of carnivrous plants, however despite early conjecture that the carnivorous syndrome was of monophyletic origin, sequence data have demonstrated that carnivores evolved independently from various angiosperm lineages (Albert et al., 1992; Ellison and Gotelli, 2001; Heubl et al., 2006; Ellison and Gotelii, 2009).

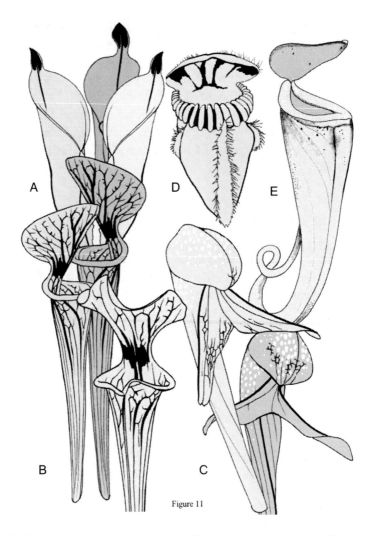

Figure 5. The three independent angiosperm lineages of pitchers, in the Ericales: A. *Heliamphora*; B. *Sarracenia* and C. *Darlingtonia*, the Oxidales: D. *Cephalotus* and the Caryophyllales: E. *Nepenthes*.

Phylogenetic analyses revealed that the pitcher plants belonging to the families Nepenthaceae, Sarraceniaceae and Cephalotaceae (Figure 11) were in fact only distinctly related, even though two independently arose from flypaper lineages (Albert et al., 1992). New molecular techniques that bring together evolutionary and developmental biology ('evo-devo'), have the potential to offer interesting insights into the evolution of *Nepenthes* and other carnivorous plants.

CONCLUSION

Extreme divergence in morphological features associated with the carnivorous syndrome, coupled with extraordinary degrees of convergence obscured the evolutionary history of carnivorous plants including *Nepenthes*. Early studies on the evolution of carnivorous plants were therefore largely speculative and unsubstantiated.

Advances in molecular phylogenetics towards the end of the 20th Century have vastly improved our understanding of the evolution of flowering plants, and revealed the origin of carnivory within the Caryophyllales - a diverse order of core eudicots that includes succulents, cacti and plants with specialised mechanisms of nutrient uptake, which probably predisposed the order to carnivory.

Combined gene phylogenetic analyses and morphological data have provided strong evidence for the monophyly of the carnivorous clade in the Caryophyllales, which offers a platform for generating hypotheses about the evolution of carnivory in *Nepenthes* and related taxa. Importantly, evidence now suggests that the ancestral state linking all carnivores was most probably a flypaper trap similar to those of the extant Drosophyllaceae and Droseraceae (Meimberg et al., 2000). The fossil record of carnivorous angiosperms is depauperate and fragmentary, therefore to understand the evolutionary shift to carnivory, scientists have looked to extant examples of plants that may be in the transition from non-carnivory to carnivory – the protocarnivores.

Such plants may provide a framework for testing hypotheses and gathering empirical data on the evolution of carnivory in the plant kingdom. Taken together, it is likely that during the evolution of the *Nepenthes* pitcher, leaf transformations caused the embedding and subsequent partial covering of glands by epidermal tissue in the putative Droseraceae-like progenitor (Meimberg et al., 2000; Heubl et al., 2006).

It is suggested that new molecular techniques that bring together evolutionary and developmental biology ('evo-devo'), have the potential to offer exciting insights into the evolution of *Nepenthes* and other carnivorous plants.

REFERENCES

Adam, JH, Wilcock CC, 1991. A new species of *Nepenthes* (Nepenthaceae) from Sarawak. *Blumea* 36: 123–125. Adamec L, 1997. Mineral nutrition of carnivorous plants: a review. *The Botanical Review* 63:273–299.

Airy Shaw HK, 1951. On the Dioncophyllaceae, a Remarkable New Family of Flowering Plants. *Kew Bulletin* 6: 327-347. Albert VA, Williams SE, Chase MW, 1992. Carnivorous Plants: Phylogeny and Structural Evolution. *Science* 257: 1491-1495. Albert VA, Stevenson DWM, 1996. Morphological cladistics of the Nepenthales. *American Journal of Botany* 83: 135 (abstract).

Anderson JAR, Muller J, 1975. Palynological study of a Holocene peat and a Miocene coal deposit from NW Borneo. *Review of Palaeobotany and Palynology* 19: 291–351. APG: Angiosperm Phylogeny Group, 1998. An ordinal classification for the families of flowering plants. *Annals of the Missouri Botanical Garden* 85: 531-553. Bailey CD, Hughes CE, Harris SA, 2004. Using RAPDs to identify DNA sequence loci for species level phylogeny reconstruction: an example from *Leucaena* (Fabaceae). *Systematic Botany* 29: 4-14. Baillon H, 1890. Observations sur quelques nouveaux types du Congo, in Bull. Soc. Linn. Paris 109: 870. Barber JT, 1978. *Capsella bursa-pastoris* seeds. Are they carnivorous? *Carnivorous Plant Newsletter* 7: 39-42.

Bauer U, Bohn HF, Federle W, 2008. Harmless nectar source or deadly trap: *Nepenthes* pitchers are activated by rain, condensation and nectar. *Proceedings of the Royal Society B* 275: 259-265.

Bohn HF, Federle W, 2004. Insect aquaplaning: *Nepenthes* pitcher plants capture prey with the peristome, a fully wettable water-lubricated anisotropic surface. *Proceedings of the National Academy of Sciences* 101: 14138–14143. Breuker CJ, Debat V, Klingenberg CP, 2006. Functional evo-devo. *Trends in Ecology and Evolution* 21: 488-492. Cameron KM, Kenneth J, Wurdack KJ, Jobson RW, 2002. Molecular evidence for the common origin of snap-traps among carnivorous plants. *American Journal of Botany* 89: 1503–1509..

Cheek M, Jebb M, Lee CC, Lamb A, Phillipps A, 2003. *Nepenthes hurrelliana* (Nepenthaceae), a new species of pitcher plant from Borneo. *Sabah Parks Nature Journal* 6: 117–124. Christy M, 1923. The common teasel as a carnivorous plant. *Journal of Botany* 61: 33-45.

Clarke CM, 1997. *Nepenthes of Borneo*. Natural History Publications (Borneo), Kota Kinabalu.

Clarke CM, 1999. *Nepenthes benstonei* (Nepenthaceae), a new pitcher plant from Peninsular Malaysia. *Sandakania* 13: 79–87.

The Evolution of Carnivory in Flowering Plants   173

Clarke CM, 2001a. A guide to the pitcher plants of Sabah. *Natural History Publications (Borneo)*, Kota Kinabalu.

Clarke CM, 2001b. *Nepenthes* of Sumatra and Peninsular Malaysia. *Natural History Publications (Borneo)*, Kota Kinabalu.

Clarke CM, 2002. A guide to the pitcher plants of Peninsular Malaysia. *Natural History Publications (Borneo)*, Kota Kinabalu. Clarke CM, Kitching RL, 1995. Swimming ants and pitcher plants: a unique ant-plant interaction from Borneo. *Journal of Tropical Ecology* 11: 589-602.

Clarke CM, Lee CC, 2004. A pocket guide: pitcher plants of Sarawak. *Natural History Publications (Borneo)*, Kota Kinabalu.

Clarke CM, Lee CC, McPherson S, 2006. *Nepenthes chaniana* (Nepenthaceae), a new species from north-western Borneo. *Sabah Parks Journal* 7: 53–66. Cronquist, 1988. The evolution and classification of flowering plants. *Columbia University Press*, New York. Croizat L. 1960. *Principia Botanica*, or beginnings of botany (with sketches by the author). Caracas, Published by the Author.

Cuénoud P, Savolainen V, Chatrou LW, Powell M, Grayer RJ, Chase MW, 2002. Molecular Phylogenetics of Caryophyllales Based on Nuclear 18S rDNA and Plastid *rbc*L, *atp*B, and *mat*K DNA Sequences. *American Journal of Botany* 89: 132-144.

Danser BH, 1928. The Nepenthaceae of the Netherlands Indies. *Bulletin de Jardin de Botanique*, Buitenzorg,. 9: 249-438.

Darnowski DW, Carroll DM, Płachno B, Kabanoff E, Cinnamon E, 2006. Evidence of Protocarnivory in Triggerplants (*Stylidium* spp.; Stylidiaceae). *Plant Biology* 8: 805–812. Darwin C. 1859. The origin of species by means of natural selection. *Oxford University Press,* Oxford. Darwin C. 1875. Insectivorous plants. *John Murray*, London. Davis CC, Latvis M, Nickrent DL, Wurdack KJ, Baum DA, 2007. Floral gigantism in Rafflesiaceae. *Science* 315: 1812.

de Flacourt E. 1658. *Histoire de la Grande Isle de Madagascar*. L'Amy, Paris.

Downie SR, Katz-Downie DS, Kyung-Jin C, 1997. Relationships in the Caryophyllales as suggested by phylogenetic analyses of partial chloroplast DNA ORF2280 homolog sequences. *American Journal of Botany* 84: 253-273. Ellison AM, Gotelli NJ, 2001. Evolutionary ecology of carnivorous plants. *Trends in Ecology and Evolution* 16, 623–629. Ellison AM, Gotelli NJ, 2009. Energetics and the evolution of carnivorous plants—Darwin's _most wonderful plants in the world'. *Journal of Experimental Botany* 60: 19–42. Engler A, 1898. Syllabus der pflanzenfamilien. Verlag von Gebruder Borntraeger, Berlin. Fay MF, Cameron KM, Prance GT, Lledó MD, Chase MW, 1997. Familial Relationships of *Rhabdodendron* (Rhabdodendraceae): Plastid *rbc*L Sequences Indicate a Caryophyllid Placement. *Kew Bulletin* 52: 923-932.

174   Chris Thorogood

Frazier CK, 2000. The enduring controversies concerning the process of protein digestion in *Nepenthes* (Nepenthaceae). *Carnivorous Plant Newsletter* 29: 56–61.

Gaume L, Gorb S, Rowe N, 2002. Function of epidermal surfaces in the trapping efficiency of *Nepenthes alata* pitchers. *New Phytologist* 156, 479–489.

di Guisto B, Grosbois V, Fargeas E, Marshall DJ, Gaume L, 2008. Contribution of pitcher fragrance and fluid viscosity to high prey diversity in a *Nepenthes* carnivorous plant from Borneo. *Journal of Bioscience* 33: 121–136.

Gilbert LE, 1971. Butterfly-plant coevolution: Has *Passiflora* won the selectional race with Heliconiine butterflies? *Science* 172: 585-586.

Givnish TJ, Burkhardt EL, Happel RE, Weintraub JD, 1984. Carnivory in the bromeliad *Brocchinia reducta*, with a cost/benefit model for the general restriction of carnivorous plants to sunny, moist, nutrient-poor habitats. *The American Naturalist* 124: 479-497. Goodman CS, Coughlin BS, 2000. The evolution of evo-devo biology. *Proceedings of the National Academy of Sciences* 97: 4424–4456.

Green S, Green TL, Heslop-Harrison, 1979. Seasonal heterophylly and leaf gland features in *Triphyophyllum* (Dioncophyllaceae), a new carnivorous plant genus. *Botanical Journal of the Linnean Society* 78: 99-116.

Hanson AD, Rathinassabapathi B, Rivoal J, Burnet M, Dillon MO, Gage DA, 1994. Osmoprotective compounds in the Plumbaginaceae: a natural experiment in the metabolic engineering of stress tolerance. *Proceedings of the Natural Academy of Sciences* 91: 306-310.

Harborne JB, 1967. Comparative biochemistry of the flavenoids-IV. Correlations between chemistry, pollen morphology and systematics in the family Plumbaginaceae. *Phytochemistry* 6: 1415-1428. Hartmeyer S, 1998. Carnivory in *Byblis* revisited II: The phenomenon of symbiosis on insect trapping plants. *Carnivorous Plants Newsletter* 27, 110-113. Heads M, 1984. *Principia Botanica*: Croizat's contribution to Botany. *Tuatara* 27: 26-48.

Heubl GR, Wistuba A, 1995. A cytological study of the genus *Nepenthes* L. (Nepenthaceae). *Sendtnera* 4:169–174. Heubl G, Bringmann G, Meimberg H, 2006. Molecular phylogeny and character evolution of carnivorous plant families in caryophyllales – Revisited. *Plant Biology* 8: 821–830.

Hooker JD, 1859. On the origin and the development of the pitchers of *Nepenthes*, with an account of some new Bornean plants of that genus. *The Transactions of the Linnean Society of London* 22: 415–424.

Hooker JD, 1873. Nepenthaceae. *In*: De Candolle A, *Prodromus systematis universalis regni vegetabilis* 17: 90–105. Hurlbert SH, 1971. The non-concept of species diversity: a critique and alternative parameters. *Ecology* 52: 577–586.

The Evolution of Carnivory in Flowering Plants   175

Jaffe K, Michelangeli F, Gonzalez JM, Miras B, Ruiz MC, 1992. Carnivory in pitcher plants of the genus *Heliamphora* (Sarraceniaceae). *New Phytologist.* 122: 733-744.

Jebb M, Cheek M, 1997. A skeletal revision of *Nepenthes* (Nepenthaceae). *Blumea* 42: 1-106. Judd WS, Campbell CS, Kellogg EA, Stevens PE, 1999. Plant systematics: a phylogenetic approach. *Sinauer,* Sunderland, Massachusetts. Juniper BE, Robins RJ, Joel DM, 1989. The carnivorous plants. *Academic Press Limited*, London.

Korthals PW, 1839. Over het geslacht *Nepenthes. In* Temminck CJ, Verhandelingen over de natuurlijke geschiedenis der Nederlandsche overzeesche bezittingen; Kruidkunde. 1-44.

Krutzsch W, 1985. Über Nepenthes-Pollen im europäischen Tertiär. *Gleditschia* 13: 89-93. Krutzsch W, 1988. Palaeogeography and historical phytogeography (palaeochorology) in the Neophyticum. *Plant Systematics and Evolution* 162: 5-61.

Linnaeus C. 1737. *Nepenthes. Hortus Cliffortianus.* Amsterdam.

Linnaeus C. 1753. *Nepenthes. Species Plantarum* 2: 955.

Lledó MD, Crespo MB, Cameron KM, Fay MF, Chase MW, 1998. Systematics of Plumbaginaceae Based upon Cladistic Analysis of *rbc*L Sequence Data. *Systematic Botany* 23: 21-29.

Mameli E, 1916. Richerche anatomiche, fisiologiche e biologiche sulla *Martynia lutea* Lindl. Att. Ist. Bot. Dell'Universita di Pavia 16: 137-188.

Mameli E, Aschieri E, 1920. Ricerche anatomiche, e biomiche sul *Lychnis viscaria.* Att. Ist. Bot. Dell'Universita di Pavia, 17: 119-129.

Marburger JE, 1979. Glandular Leaf Structure of *Triphyophyllum peltatum* (Dioncophyllaceae): A "Fly-Paper" Insect Trapper. *American Journal of Botany* 66: 404-411. Meimberg H, Dittrich P, Bringmann G, Schlauer J, Heubl G, 2000. Molecular phylogeny of Caryophyllidae s. l. based on *mat*K- sequences with special emphasis on carnivorous taxa. *Plant Biology* 2: 218-228.

Meimberg, H Wistuba A, Dittrich P, Heubl G, 2001. Molecular Phylogeny of Nepenthaceae Based on Cladistic Analysis of Plastid *trn*K Intron Sequence Data. *Plant biology* 3: 164-175.

Meimberg H, Heubl G, 2006. Introduction of a Nuclear Marker for Phylogenetic Analysis of Nepenthaceae. *Plant Biology* 8: 831–840.

Moran JA, 1996. Pitcher Dimorphism, Prey Composition and the Mechanisms of Prey Attraction in the Pitcher Plant *Nepenthes rafflesiana* in Borneo. *Journal of Ecology* 84: 515-525.

Moran JA, Booth WE, Charles JK, 1999. Aspects of pitcher morphology and spectral characteristics of six Bornean *Nepenthes* pitcher plant species: implications for prey capture. *Annals of Botany* 83, 521–528

176 Chris Thorogood

Moran JA, Moran AJ 1998. Foliar Reflectance and Vector Analysis Reveal Nutrient Stress in Prey-Deprived Pitcher Plants (*Nepenthes rafflesiana*). *International Journal of Plant Sciences* 159: 996–1001.

Moran JA, Merbach MA, Livingston NJ, Clarke CM, Booth WE, 2001. Termite prey specialization in the pitcher plant *Nepenthes albomarginata* - evidence from stable isotope analysis. *Annals of Botany* 88: 307-311.

Moran JA, Clarke CM, Hawkins BJ, 2003. From Carnivore to Detritivore? Isotopic evidence for leaf litter utilization by the tropical pitcher plant *Nepenthes ampullaria*. *International Journal of Plant Sciences* 164: 635–639. Muller KF, Borsch T, Legendre L, Porembski S, Barthlott W, 2006. Recent progress in understanding the evolution of carnivorous Lentibulariaceae (Lamiales). *Plant Biology* 8, 748–757.

Nakayama S, Amagase S, 1968, Acid protease in *Nepenthes*: Partial purification and properties of the enzyme. *Proceedings of the Japan Academy* 44: 358-362.

Phillipps A, Lamb A, 1996. Pitcher plants of Borneo. *Natural History Publications (Borneo),* Kota Kinabalu.

Phillipps A, Lamb A, Lee CC, 2008. Pitcher plants of Borneo, second edition. *Natural History Publications (Borneo),* Kota Kinabalu. Poiret JLM,1797. Népente. *In*: Lamarck JB, Encyclopédie Méthodique Botanique: 4. Rachmilevitz T, Joel DM, 1976. Ultrastructure of the calyx glands of *Plumbago capensis* Thunb., in relation to the process of secretion. *Israel Journal of Botany* 25: 127 - 139.

Radhamani TR, Sudarshana L, Krishnan R, 1995. Defense and carnivory: Dual role of bracts in *Passiflora foetida*. *Journal of Biosciences* 20: 657-664. Raven PH, Axelrod DJ, 1974. Angiosperm biogeography and past continental movements. *Annals of the Missouri Botanical Garden* 61: 539-673. Savolainen V, Fay MF, Albach DC, Backlund A, van der Bank M, Cameron KM, Johnson SA, Lledó MD, Pintaud J-C, Powell M, Sheahan MC, Soltis DE, Soltis PS, Weston P, Whitten WM, Wurdack KJ, Chase MW, 2000. Phylogeny of the eudicots: a nearly complete familial analysis based on *rbc*L gene sequences. *Kew Bulletin* 55: 257- 309. Schlauer J, 1996. —New‖ data relating to the evolution and phylogeny of some carnivorous plant families. *Carnivorous Plant Newsletter* 26: 34-38. Simons P, 1981. How exclusive are carnivorous plants? *Carnivorous Plant Newsletter* 10: 65 – 80. Soltis DE, Soltis PS, Nickrent DL, Johnson LA, Hahn WJ, Hoot SB, Sweere JA, Kuzoff RK, Kron RK, Chase MW, Swensen SM, Zimmer EA, Chaw S-M, Gillespie LJ, Kress WJ, Sytsma KJ, 1997. Angiosperm phylogeny inferred from 18S ribosomal DNA sequences. *Annals of the Missouri Botanical Garden* 84: 1-49.

Spomer GG, 1999. Evidence of protocarnivorous capabilities in *Geranium viscosissimum* and *Potentilla arguta* and other sticky plants. *International Journal of Plant Sciences* 160: 98-101.

The Evolution of Carnivory in Flowering Plants   177

Sprague TA, 1916. *Dioncophyllum*; with note on the anatomy by F.M. Scott; In: *Bulletin of Miscellaneous Information* 89-92. Stevens PF, 2007. Angiosperm Phylogeny Website, Version 8. http://www.mobot.org/OBOT/research/APweb. Swofford DL, 1996. PAUP*: *Phylogenetic Analysis Using Parsimony (*and Other Methods) Version 4.0.* Sinauer Associates, Sunderland, Massachusetts. Takhtajan A, 1969. Flowering plants – origin and dispersal. Smithsonian Institution Press, Washington DC. Thorne RF, 1992. Classification and Geography of the Flowering Plants. *Botanical Review* 58: 225-348. Warburg O, 1893. Flacourtiaceae, in Engler u. Prantl, Natiirl Pflanzenfam. III, 6: 29-30.

Williams SE, Albert VA, Chase MW, 1994. Relationships of Droseraceae: A Cladistic Analysis of *rbc*L Sequence and Morphological Data. *American Journal of Botany* 81: 1027-1037. Zambelli E, 1929. Ricerche anatomofisiologiche sulla *Petunia violacea* e sull *Petunia nyctaginiflora* come piante insettivore. Att. Ist. Bot. Dell'Universita di Pavia, 1: 75-87.

INDEX

#

21st century, 3, 114

A

abscisic acid, ix, 5, 71, 72, 79, 85, 86, 88, 90, 91, 92, 93, 94, 95, 96, 142
acetic acid, 86, 94
acid, ix, 5, 12, 52, 71, 72, 76, 78, 79, 83, 85, 86, 87, 88, 90, 91, 92, 93, 94, 95, 96, 122, 123, 139, 140, 142, 144, 160
acidic, 159, 164
acidity, 75
ACL, 107
adaptations, 155
adenine, 81
adverse effects, viii, 2, 25, 31, 37, 38, 79
aesthetic, 140
agriculture, 61, 79, 112
Agrobacterium, 103, 108
air temperature, vii, 2, 3, 7, 9, 13, 21, 34
airborne particles, 62
alkaloids, 49
allopolyploid, 115
alters, 110
amino acids, 75, 163, 167
amylase, 74, 80, 85, 96

anatomy, 164, 165, 166, 177
ancestors, xi, 153, 157, 159, 163
aneuploid, 109, 110
angiosperm, 154, 155, 169, 170
ANOVA, ix, 44, 59, 60
anther, 161
antioxidant, 74, 88
apex, 130
Apomixis, v, ix, 97, 98, 99, 106, 111, 112, 116, 117
apples, 51, 53, 61
Arabidopsis thaliana, 82, 85, 86, 91, 102, 115, 116, 150
arbuscular mycorrhizal fungi, 91
aromatic hydrocarbons, 69
aromatic rings, 48
arthropods, 167
assimilation, 120, 130
atmosphere, 20, 48, 62, 63, 65
atmospheric deposition, 59, 64, 68
ATP, 74, 123
auxin(s), ix, 71, 72, 80, 81, 83, 85, 86, 87, 88, 89, 93, 94, 95, 117
azaleas, 146, 147, 151

B

BAC, 113
bacteria, ix, 71, 72, 83, 84, 91

Index

bacterial artificial chromosome, 108, 109
bacterium, 103
behavioral disorders, 46
bioaccumulation, 47, 48
biochemical processes, 45, 72
biochemistry, 87, 174
biodegradation, 48
biodiversity, 87, 98
biogeography, 177
biomass, 3, 5, 31, 39, 146
biosynthesis, 75, 79, 85, 86, 89, 92, 96, 143, 145, 151
biotechnology, 110, 112, 114
body weight, 62
boll production, vii, 1, 3, 5, 9, 10, 12, 13, 14, 15, 16, 17, 20, 21, 24, 25, 27, 29, 30, 32, 34, 35, 37, 41
boll retention patterns, vii, 1
boll setting, vii, 1, 4, 10, 20
bone marrow, 45
branching, 139, 148, 151
brassinosteroids, ix, 71, 72, 76, 84, 88, 89, 96
breeding, vii, ix, 3, 98, 100, 106, 110, 112, 156
budding, 20
bursa, 172

C

Ca^{2+}, 72
cabbage, 88, 94
cadmium, 62, 66
calcium, 6, 45, 46, 144, 152
calcium carbonate, 6
calibration, 54, 55, 58
calyx, 51, 160, 176
carbohydrate(s), x, 119, 120, 124, 125, 126, 127, 128, 129, 130, 131, 132, 133, 134
carbohydrate metabolism, x, 120, 130, 131, 132
carbon, 37, 38, 41, 82, 133, 135
carbon dioxide, 41
carboxylic acid, 77, 84, 88, 92
carcinogen, 45
carnivores, 155, 159, 160, 163, 165, 169, 171

Caucasus, 51
cDNA, 90
cell cycle, 74, 75, 103
cell division, x, 73, 74, 91, 98, 109, 110
cell surface, 82
cellulose, 39
central nervous system, 45
centromere, 107, 108, 109, 110, 114, 116
chemical, 44, 48, 61, 67, 93, 133, 139, 140, 143, 151, 167
chemical properties, 61
chemicals, 44, 51, 63, 72, 102, 139, 140, 143, 144, 146, 166
chlorophyll, 134, 140, 144
chloroplast, 77, 173
chromatid, 105
chromatography, 54
chromosome, 99, 103, 106, 107, 108, 109, 113, 114, 116
classification, vii, 155, 156, 172, 173
climate, 3, 16, 17, 39, 41, 142
climate change, 3, 41
climates, 41
climatic factors, vii, 1, 3, 5, 7, 8, 10, 11, 13, 16, 17, 18, 19, 20, 21, 22, 23, 24, 25, 27, 28, 29, 30, 32, 33, 34, 35, 36, 37, 38, 41
clone, 104
cloning, 101
closure, 31
CO2, 4, 20, 40, 98, 120, 130
coal, 48, 65, 172
coal tar, 65
cobalt, 104
codon, 111
commercial crop, 101, 112
composition, 99, 104, 106, 110, 133
compounds, 47, 48, 63, 74, 75, 89, 138, 140, 144, 163, 174
conductance, 5, 42, 134
consensus, 138
conservation, 87
consumers, 66, 110
consumption, viii, 43, 45, 62, 63, 151
containers, x, 137, 143
contaminated food, 45

Index

181

contaminated soil, 45, 65, 69
contaminated water, 45
contamination, 45, 59, 62, 66, 122
contradiction, 26, 37, 130
convention, 47, 67
convergence, 155, 156, 171
correlation analysis, 5
correlation coefficient, 9, 18, 19, 22, 23, 32, 33, 34, 36
cotton, vii, 1, 3, 4, 5, 6, 8, 9, 10, 12, 13, 15, 16, 24, 25, 27, 29, 30, 31, 34, 35, 36, 37, 38, 39, 40, 41, 42
cotton cultivars, vii, 1, 15, 31
cotyledon, 80
crop, vii, 1, 2, 4, 5, 6, 15, 31, 84, 103, 107, 114, 116, 134, 138, 140, 142
crop production, 2, 4, 84
crops, x, 3, 39, 41, 59, 86, 89, 98, 99, 103, 106, 111, 115, 119, 120, 138, 146, 167
crown, 51, 147
crystal structure, 94
cultivars, vii, x, 1, 3, 4, 15, 31, 39, 45, 87, 120, 123, 130, 131, 132, 134, 135, 137, 139, 141, 142, 143, 144, 145, 146, 147, 151, 152
cultivation, 147
cultural practices, viii, 2, 3, 5, 38
culture, 92, 111, 114, 139
culture medium, 92
Cydonia oblonga, 51
cysteine, 78, 85
cytochrome, 144
cytogenetics, 113
cytokinins, ix, 71, 72, 76, 81, 82, 83, 84, 92, 93, 134, 142, 151
cytometry, 104, 106

D

deoxyribonucleic acid, 91
dependent variable, 7, 8, 11, 12, 13, 16, 24, 30, 36
derivatives, 44, 82
desiccation, 73, 74
developmental process, 4

dew, 3
digestive enzymes, 166
dimethylsulfoxide, 123
diploid, 99, 100, 101, 102, 104, 109, 110
diseases, 44, 45, 46, 139
dispersion, 160
dissociation, 82, 95
distilled water, 5, 123
distribution, 44, 47, 59, 66, 69, 131, 132, 133, 135
divergence, x, 153, 171
diversification, 162
diversity, 174, 175
DNA, ix, 73, 83, 97, 100, 101, 102, 103, 104, 105, 107, 108, 109, 113, 114, 116, 158, 172, 173, 177
drought, 15, 31, 82, 94, 160
dry matter, 12, 133
drying, 123

E

ecology, 173
economical crop management, vii, 1
ecosystem, 45
egg, ix, 98, 99, 100, 111
Egyptian cotton, vii, 1, 5, 8, 9, 10, 13, 24, 25, 30, 34, 37, 41
electron, ix, 44, 56
elongation, 73, 91, 96, 138, 139, 140, 142, 143, 144, 146, 150
embryo sac, 102, 113
embryogenesis, 73, 76, 82, 86, 87, 89, 90, 91, 99, 103, 117
endocrine disorders, 46
endosperm, x, 75, 77, 78, 79, 80, 82, 83, 91, 98, 99, 100, 101, 102, 103, 104, 105, 106, 107, 113, 117, 159
engineering, 103, 106, 107, 108, 109, 111, 113, 115, 174
environment, 3, 5, 8, 44, 45, 47, 48, 61, 62, 65, 66, 67, 112, 117
environmental change, 98
environmental conditions, 2, 31, 120, 121, 140, 143

182 Index

environmental factors, 2, 139
environmental stress, 131
environmental stresses, 131
enzyme(s), 4, 45, 46, 73, 74, 75, 77, 78, 81, 82, 83, 85, 91, 130, 131, 133, 167, 176
EPA, 48, 54, 67
epigenetics, 115
ethanol, 123
ethylene, ix, 71, 72, 75, 76, 77, 78, 79, 81, 83, 84, 85, 86, 90, 91, 92, 93, 96
evaporation, viii, 2, 5, 7, 9, 10, 12, 13, 17, 19, 20, 21, 24, 25, 27, 30, 31, 34, 35, 37
evapotranspiration, 15, 41
evergreen azalea, x, 137
evolution, vii, xi, 101, 133, 153, 155, 156, 159, 160, 161, 163, 165, 168, 169, 171, 173, 174, 175, 176, 177
experimental condition, 121
experimental design, 121
exploitation, 158
exposure, viii, 4, 15, 43, 44, 45, 46, 48, 67, 102, 167
expressed sequence tag, 103, 117
extraction, 54, 57, 72
extracts, 54
exudate, 150

F

farmers, 99, 116
farmland, 42
fatty acids, 140
fermentation, 12
fertility, 3, 111, 133, 138
fertilization, ix, 15, 31, 97, 99, 101, 115, 116, 141
fertilizers, 62, 139, 140, 151
fiber(s), 3, 4, 39, 40, 41
field trials, 6, 12, 31
filiform, 161, 164
fish, 47
fishing, 53
fixation, 74, 99, 102, 116, 160
flowering period, 7, 147

flowers, 4, 5, 6, 7, 8, 9, 12, 13, 14, 16, 24, 25, 26, 35, 50, 51, 120, 140, 147, 156, 167
fluctuations, 21, 35, 38, 125, 131
fluorescence, 84, 134
food, viii, 39, 43, 44, 45, 46, 47, 49, 50, 57, 62, 64, 65, 66, 68, 69, 98, 111, 113, 169
food additives, 44
food chain, 47
food production, 98
forecasting, 21, 38
forest fire, 48
fructose, 123, 125, 126, 127, 128, 129, 130, 131
Fruiting of cotton plant, vii, 1
fruits, viii, 44, 49, 51, 52, 53, 61, 62, 63, 64, 65, 66, 67, 69, 88, 167
functional analysis, 88
fungus, 79
fusion, 100, 101, 102, 162, 168, 169

G

gamete, 99, 103
gametophyte, ix, 97, 112
gene expression, 85, 96, 104, 106, 116
genes, ix, 76, 77, 78, 79, 80, 81, 83, 86, 87, 88, 90, 91, 98, 100, 101, 103, 106, 107, 108, 109, 110, 117, 163
genetics, 91, 113, 114, 115
genome, 102, 103, 104, 107, 110, 111, 113, 114, 116, 117
genomic regions, 101, 107
genotype, 99, 103
genus, x, 49, 50, 51, 104, 137, 156, 157, 163, 164, 167, 168, 174, 175
germ cells, 102
germination, vii, ix, 71, 72, 73, 74, 75, 76, 77, 78, 79, 80, 81, 82, 83, 84, 85, 86, 87, 88, 89, 90, 91, 92, 93, 94, 95, 96
gibberellin(s), ix, 71, 72, 73, 75, 76, 79, 80, 81, 83, 84, 86, 87, 88, 89, 91, 92, 94, 95, 96, 143, 150, 151
gigantism, 173
glucose, 123, 125, 126, 127, 128, 129, 131
glutamate, 87

glutamine, 75
glycine, 12, 40
grass, 7, 59
greenhouse, x, 5, 122, 137, 151
greenhouse-grown potted plant, x, 137
groundwater, 62
growth rate, 61
growth retardants, 138, 139, 143, 150, 151

H

habitat(s), 50, 157, 160, 161, 168, 174
haploid, 100, 101, 102, 110, 111, 115
harvesting, 31, 93
heavy metals, viii, 44, 45, 48, 54, 59, 60, 61, 65, 66, 67, 68, 69, 82
heavy particle, 48
height, 50, 106, 138, 142
hemoglobin, 45, 46
herbicide, 108
heterochromatin, 101
heterosis, 99, 110
hexane, 54
histidine, 72, 81, 82, 95
histone, 101, 109, 110
histones, 113
Holocene, 172
homeostasis, 88
homogeneity, 55, 58, 140
homologous chromosomes, 102
honey bees, 51
hormone(s),vii, ix, 71, 72, 73, 74, 75, 76, 77, 78, 79, 80, 81, 82, 83, 84, 85, 87, 89, 90, 93, 95
humidity, vii, 2, 3, 5, 7, 10, 11, 13, 14, 15, 17, 19, 20, 21, 24, 25, 27, 29, 30, 31, 32, 34, 35, 37, 38, 42, 122
hybrid, 72, 99, 111, 114, 115, 116, 141
hybridization, 99
hydrocarbons, 69
hydrogen peroxide, 52
hydrophobicity, 47
hydroxyl, 91
hyperactivity, 46
hypothesis, 155, 159, 160, 161, 165

hypoxia, 74

I

image analysis, 138, 148, 149, 151
immune system, 46
imprinting, 101, 103, 113, 114, 115
improvements, 141
in situ hybridization, 101, 108
in vitro, 89, 92
inbreeding, 99
incompatibility, 99, 101
incomplete combustion, 48
independent variable, 7, 9, 11, 13, 16, 24, 25
induction, 138, 141, 143, 151
industrial emissions, 45
inhibition, 74, 75, 78, 91, 95, 145
inhibitor, 147
initiation, 5, 24, 25, 37, 75, 81, 105, 141, 143, 144, 151, 152
insects, 51, 160, 163, 166, 167, 168, 169
insertion, 108, 109
integration, 109
interference, 46, 111
internal mechanisms, 120
International Atomic Energy Agency, 52
internode, 142, 144
inversion, 111
ionization, 56
ions, 45, 133
iron, 46
irradiation, 104
irrigation, 6, 8, 9, 12, 15, 38, 45, 60
isotope, 38, 176

K

K^+, 72, 74

L

landfills, 62
landscape, x, 137, 143
larvae, 166

lead, 5, 20, 37, 38, 45, 66, 68, 85, 86, 101, 106, 112, 147
legume, 68
leucine, 82, 90
ligand, 81
light, 3, 4, 14, 15, 31, 48, 75, 83, 88, 92, 93, 94, 138, 142, 143, 150, 151
linear model, 7
lipid peroxidation, 85
liver, 45, 62
livestock, 47
loci, 101, 104, 106, 107, 108, 109, 110, 172
locus, 104, 107, 108, 112
low temperatures, 31, 140
Luo, 104, 105, 115

M

mammalian cells, 66
manganese, 45
mass spectrometry, 68
Mauritius, 152
megaspore, 102
meiosis, x, 99, 100, 102, 103, 107, 108, 113, 116, 120, 121, 123, 131
membrane permeability, 75
mental development, 45
meristem, 105, 140, 150
metabolic pathways, 86, 151
metabolism, 37, 45, 46, 85, 88, 93, 96, 134, 135, 144, 160
metabolites, vii, 83, 86, 91
metabolized, 146
metallurgy, 45
metals, viii, 43, 45, 52, 53, 56, 59, 60, 61, 66, 67, 68
methanol, 12, 40
methylation, 101, 105, 114
microorganisms, 72, 83
microRNA, 92
mineralization, 53
Miocene, 172
mitosis, 103, 108, 110, 113
model system, xi, 153
models, 3, 25, 26, 27, 132, 163

modifications, x, 102, 110, 112, 119, 131, 137, 153, 167
moisture, 3, 12, 40, 74
molecular biology, ix, x, 71, 73, 84, 98
molecular weight, 48
molecules, 81, 106
morphogenesis, 72, 92, 95, 151
morphology, 156, 161, 174, 176
mortality rate, 145
mother cell, 102
moving window, 17
mRNA, 106
multiple regression, 3, 24, 35
multiple regression analysis, 35
mutant, 88, 91, 106, 110
mutation, 86, 110, 111, 158, 169
mutation rate, 158
mutations, 86, 102, 169

N

Na^+, 74
naphthalene, 48
natural selection, 173
necrosis, 124, 125, 126, 127, 129, 130, 133
negative effects, viii, 2, 5, 38, 140
negative relation, 9, 10, 15, 24, 30, 31, 34, 36
neural networks, 150
neutrons, 102
nitric oxide, 88, 90, 96
nitrite, 75
nitrogen, 6, 54, 73, 75, 98, 133, 139, 140, 141, 160
nitrogen gas, 54
nitrous oxide, 74
nodes, 122
nucellus, 102
nuclei, 100, 102, 106
nucleus, 79, 81, 96
nutrient, 12, 45, 74, 84, 120, 130, 133, 139, 140, 159, 160, 161, 164, 166, 167, 168, 169, 171, 174
nutrients, 3, 120, 140, 141, 167, 169
nutrition, x, 133, 139, 141, 153, 172
nutritional status, 139, 143

Index   185

O

oil, 31
organ, x, 119, 131
organic matter, 6, 45, 53, 63
Organochlorine pesticides (OCPs), viii, 43, 44, 47, 48, 54, 56, 59, 60, 63, 64, 65, 68, 69
origins of carnivory, xi, 153, 155, 157, 163
ornamental plants, 138, 143, 147
ovaries, 50
ovule, 100, 101, 102, 104, 120, 121
oxidation, 74, 78, 144
oxidative damage, 74, 96
oxidative stress, 82, 86, 93
oxygen, 38, 46, 74, 79

P

parenchyma, 46
parthenogenesis, x, 98, 99, 100, 101, 103, 104, 107, 117
pathogens, 76, 79, 84
pathways, viii, 43, 48, 61, 72, 75, 78, 79, 80, 84, 89, 95, 115
PCBs, 47
peat, 172
peristome, 172
Persistent Organic Pollutants, viii, 43, 44
perylene, 49
pesticide, 46, 57, 68
pests, vii, 1, 3, 47, 98, 139
pH, 6, 8, 9, 45, 76
phenotype(s), 94, 103, 106, 114
phosphate, 62, 123
phosphorus, 6, 139, 140, 141, 150, 152
phosphorylation, 82
photosynthesis, 3, 12, 31, 36, 39, 40, 42, 46, 120, 121, 132, 134, 160, 163
physiology, 39, 86, 87, 90, 122, 134, 166
plant growth, 3, 4, 12, 15, 36, 38, 42, 72, 74, 77, 78, 79, 82, 83, 89, 90, 93, 95, 138, 139, 140, 141, 143, 145, 146, 149, 150, 151
plasma membrane, 76, 82, 90, 95
plastid, 91, 158

ploidy, 101, 102, 104, 110, 113, 114
polar, 90, 95, 102
polarity, 103, 116
pollen, ix, 50, 51, 80, 92, 97, 99, 100, 108, 111, 160, 174
pollen tube, ix, 97, 100
pollination, 107, 160, 167
pollinators, 51
pollutants, vii, viii, 43, 44, 47, 48, 60, 66, 67
pollution, 45, 47, 59, 65, 66
polychlorinated biphenyl, 68
polycyclic aromatic hydrocarbons (PAHs), viii, 43, 44, 48, 52, 54, 56, 57, 58, 59, 60, 64, 65, 69
polymerization, 90
polymorphism, 101
polyploid, 100, 109
positive correlation, 11, 17, 20, 21, 34, 37
potassium, 6, 140, 141
potato, 49, 50
power generation, 48
project, 3, 67, 150
proliferation, 99, 103, 105
propagation, 99, 138
propane, 92
prophase, 113
proteinase, 73, 78, 167
proteins, 73, 74, 75, 76, 78, 79, 81, 86, 87, 89, 92, 93, 104, 117
proteolytic enzyme, 95
proteomics, 104, 115
prototypes, 160
provisional tolerable weekly intake (PTWI), ix, 44, 62
public health, 45, 47
purification, 176
purity, 55

Q

quality assurance, 45
quality control, 68
quantification, 55, 57, 60, 61, 63, 64, 65, 147
quince trees, viii, 44, 49, 66

Index

R

radiation, 3, 15, 36
radicle, 73, 74, 75, 77, 78, 80, 92
rainfall, 3, 7, 8, 9, 25, 36
reactive oxygen, 79, 84
receptors, 72, 76, 82, 85, 88, 91, 92, 111
recessive allele, 103
recombinases, 111
recombination, ix, 97, 99, 100, 101, 102, 103, 104, 107, 108, 109, 114
reconstruction, 172
red blood cells, 46
regression, 9, 12, 13, 24, 25, 28, 32, 35, 55, 57, 58
regression equation, 28, 32, 35, 55, 58
regression model, 24, 25, 32
regulatory changes, 99
reproduction, x, 25, 98, 121
reproductive organs, 4, 114, 120, 129, 131
researchers, 26, 35, 37, 72, 73, 74, 77, 163
reserves, 95, 120, 121, 122, 131, 132, 133, 134, 135
residues, 53, 54, 57, 64, 66, 68
resistance, 86, 91, 104, 108, 144
resolution, 94, 112, 158, 159, 165
resources, 31, 38, 120
respiration, 74
response, vii, 1, 2, 31, 40, 42, 77, 81, 87, 93, 96, 133, 139, 149, 150
restriction fragment length polymorphis, 101
Rhododendron L. sp, x, 137
rings, 48, 82
risk assessment, vii, viii, 43, 67, 68
RNA, 111, 115
RNAi, 111
root growth, 87, 125, 141
root system, 65, 161
root(s), viii, x, 12, 44, 49, 50, 52, 59, 63, 64, 65, 66, 69, 78, 80, 81, 86, 87, 90, 93, 120, 121, 123, 124, 125, 128, 129, 130, 131, 132, 133, 134, 141, 161, 163
rural areas, viii, 44, 49, 53, 59, 66

S

salinity, 39, 75, 77, 82, 85, 89, 96
science, 67
sclerenchyma, 166
seasonality, 149
seed, vii, ix, 15, 20, 51, 71, 72, 73, 74, 75, 76, 77, 78, 79, 80, 81, 82, 83, 84, 85, 86, 87, 88, 89, 90, 91, 92, 93, 95, 96, 98, 99, 100, 101, 103, 104, 106, 113, 115, 117, 121, 160
Seed germination, ix, 71, 73, 84, 85, 86
seedling development, 88
seedlings, 79, 80, 85, 90, 91, 94, 151
segregation, ix, 98, 99, 107, 109
senescence, 81, 93, 134
sensitivity, 81, 85, 120, 121, 131, 132
sensors, 90
sepal, 51
sequencing, xi, 104, 153, 164
sexual development, 100
sexual reproduction, ix, 98, 99, 115
shape, x, 8, 112, 137, 139, 144, 147, 163
shoot(s), 63, 73, 93, 105, 130, 138, 139, 140, 141, 144, 148, 150, 164
shrubs, x, 50, 137
signal transduction, 88, 90, 94
signaling pathway, 81, 83
silica, 52, 54, 56
skin, 46
sludge, 62
sodium, 52, 54
soil particles, 63
soil pollution, 64
solubility, 48
solution, 12, 54, 57, 123, 141
sorption, 63
sowing, 15, 31
soybeans, 134
species, x, 15, 45, 49, 50, 51, 61, 78, 79, 81, 84, 87, 96, 99, 107, 110, 114, 116, 130, 137, 138, 139, 140, 141, 142, 143, 146, 149, 150, 153, 155, 156, 157, 159, 160, 163, 164, 165, 166, 167, 168, 169, 172, 173, 175, 176
speculation, 49, 155, 156

Index

sperm, ix, 97, 100, 101, 102, 106, 108
spindle, 109, 110
stamens, 50, 146
starch, 120, 121, 123, 125, 126, 127, 128, 129, 130, 131, 132, 133, 134, 135, 161
sterile, 163, 165
stomata, 31
storage, 73, 74, 133, 134, 162, 169
stress, 4, 5, 12, 15, 20, 24, 25, 31, 35, 37, 39, 40, 42, 74, 75, 76, 78, 84, 85, 87, 88, 89, 90, 91, 92, 93, 94, 132, 133, 139, 140, 142, 149, 150, 160, 174
studied metals, ix, 44, 59
substrate(s), 82, 94, 144, 157
sucrose, 125, 126, 127, 128, 129, 130, 131, 132, 135
sugar beet, 88, 115
sulfate, 54
sulfur dioxide, 94
Sun, 88, 93, 113
sunshine duration, vii, 2, 7, 10, 14, 15, 17, 19, 20, 21, 24, 25, 27, 34, 37
suppression, 46, 102, 151
symbiosis, 174
symptoms, 45, 46, 140
syndrome, 155, 158, 160, 163, 165, 167, 169, 171
synergistic effect, 142
synthesis, 39, 73, 80, 89, 92, 95, 131

T

tachycardia, 46
taxa, 155, 156, 157, 158, 161, 165, 171, 176
techniques, 15, 108, 139, 155, 159, 170, 171
technologies, 100, 106, 111
technology, xi, 108, 109, 111, 116, 153
telomere, 108
temperature, vii, 2, 3, 4, 5, 7, 9, 10, 11, 13, 14, 15, 17, 20, 21, 25, 27, 28, 29, 30, 31, 34, 36, 37, 39, 40, 41, 42, 56, 75, 132, 139, 142, 143, 149, 150, 151
texture, 6, 8, 9
thinning, 6, 31, 135
tissue, 65, 77, 111, 140, 146, 162, 169, 171

topology, 158
total product, 27
toxic metals, 44, 45, 59
toxic substances, 44
toxicity, 45, 46, 48, 66
trace elements, 68
traits, ix, xi, 97, 100, 104, 107, 108, 117, 153, 155, 156, 160, 161, 164, 167
transgene, 107
transpiration, 3, 5, 20, 39, 46, 63, 65, 132
transplant, 150
transport, 42, 46, 47, 48, 63, 80, 89, 90, 93, 95, 131
treatment, viii, 43, 47, 93, 94, 122, 125, 127, 128, 130, 131, 146, 149
triploid, 100

U

urban, viii, 44, 49, 52, 53, 59, 60, 61, 63, 64, 65, 66

V

variables, 3, 11, 12, 13, 17, 20, 27, 30, 31, 32, 33, 34, 35, 36, 143
variations, x, 14, 29, 30, 61, 82, 120, 124, 125, 131, 132
vascular bundle, 164
vector, 108, 109, 111
vegetables, 45, 59, 63, 64, 65, 66, 68, 69
vegetation, 47
versatility, 47
viruses, 99
viscosity, 174
vision, 46
vitamin D, 45
volatility, 48
volatilization, 63

W

waste, 60, 61
waste disposal, 61

wastewater, 60, 62, 68
water supplies, 20, 166
WHO, 62, 68
wild type, 86, 103, 107
wildlife, 47
wood, 48, 122, 132
wool, 54

X

xylem, 133

Y

yeast, 168

Z

zinc, 46, 62, 103, 105, 109, 116
zygote, 87, 110